Also available from ASQ Quality Press

Eight-Step Process to Successful ISO 9000 Implementation:
A Quality Management System Approach
Lawrence A. Wilson

Integrating QS-9000 with Your Automotive Quality System,
Second Edition
D. H. Stamatis

QS-9000 Pioneers:
Registered Companies Share Their Strategies for Success
Subir Chowdhury and Ken Zimmer

After the Quality Audit: Closing the Loop on the Audit Process
J. P. Russell and Terry Regel

Managing Records for ISO 9000 Compliance
Eugenia K. Brumm

ISO 9000 Implementation for Small Business
James L. Lamprecht

The ISO 9000 Auditor's Companion and *The Audit Kit*
Kent A. Keeney

LearnerFirst™ How to Implement ISO 9000 software
with Dr. Lawrence A. Wilson

To request a complimentary catalog of publications,
call 800-248-1946.

Meet the Registrar

Firsthand Accounts of ISO 9000 Success from the Registration Source

C. Michael Taylor

ASQ Quality Press
Milwaukee, Wisconsin

Meet the Registrar: Firsthand Accounts of ISO 9000 Success from the Registration Source
C. Michael Taylor

Library of Congress Cataloging-in-Publication Data
Taylor, C. Michael, 1948–
    Meet the registrar: firsthand accounts of ISO 9000 success from
the registration source / C. Michael Taylor.
      p.   cm.
    Includes index.
    ISBN 0-87389-424-3 (alk. paper)
    1. ISO 9000 Series Standards. 2. Quality assurance—Management.
  I. Title.
    TS156.6.T4   1997
    658.5'62—dc21   96-50961
                                         CIP

10 9 8 7 6 5 4 3 2

ISBN 0-87389-424-3

Acquisitions Editor: Roger Holloway
Project Editor: Jeanne W. Bohn

ASQ Mission: To facilitate continuous improvement and increase customer satisfaction by identifying, communicating, and promoting the use of quality principles, concepts, and technologies; and thereby be recognized throughout the world as the leading authority on, and champion for, quality.

Attention: Schools and Corporations
ASQ Quality Press books, audiotapes, videotapes, and software are available at quantity discounts with bulk purchases for business, educational, or instructional use. For information, please contact ASQ Quality Press at 800-248-1946, or write to ASQ Quality Press, P.O. Box 3005, Milwaukee, WI 53201-3005.

For a free copy of the ASQ Quality Press Publications Catalog, including ASQ membership information, call 800-248-1946.

Printed in the United States of America

 Printed on acid-free paper

American Society for Quality

∎ASQ™

Quality Press
611 East Wisconsin Avenue
P.O. Box 3005
Milwaukee, Wisconsin 53201-3005

To my dearest Adrienne:
For 27 years
you have been
a faithful friend
and
loving, devoted wife

Contents

Foreword

What does a registrar look for when auditing a company to one of the ISO 9000 series standards or the QS-9000 requirements? The answer is simply "compliance."

Webster's dictionary defines compliance as "conformity in fulfilling official requirements." The requirements to be fulfilled in the ISO 9000 series standards are defined in eight pages and more than 2000 words. How these requirements are understood and interpreted by you and your registrar plays a significant part in your journey to certification.

This book offers an insight to the thoughts and ideas behind the reason for the creation of the standard by the International Organization for Standardization. Mike Taylor, in this publication, has presented thought-provoking information in an attempt to stretch your capacity to visualize what it takes to proceed through the certification process. He has touched upon the role and authority of the management representative, what registrars are looking for during an audit, training, quality planning, management reviews, contract reviews, document and data control, quality manual reviews, corrective action responses, and administrative process audits. Though there are many other aspects of the standard that require detailed attention, these are the building blocks for an effectively implemented system.

Mike's knowledge on the subject of quality systems is demonstrated in this collection of presentations. There is no necessity to read between the lines; he spells out his thoughts and challenges you to think of a myriad of questions even beyond those that he presents for consideration. I have attended several of Mike's presentations and have gained immensely from the material presented. I am pleased to see that he has now given many others the opportunity to learn from his experiences and knowledge.

When my company was looking for a registrar, we looked for one that we thought would be the toughest. We wanted to have the certification of our facilities mean that we have passed the ultimate test and that we have enhanced our own systems to the point that we could add to our past performance and proceed forward to our goal of becoming a world-class company. We chose our registrar because we felt confident that we would be able to work with it in a partnership to enhance our quality system to the level of certification with meaning. We have not been disappointed in the least.

Management commitment—not to the certificate on the wall, but to the effort involved in obtaining and retaining certification for the purpose of improving the processes—is of utmost importance. Without support from the top, the effort will be strenuous and meaningless. Document, implement, and check the effectiveness of the system. These three elements have been our driving force since first being involved in the ISO 9000 implementation process.

Mike takes those three elements, then expands upon them and fills in the holes in an effort to make you think beyond the general intention of the standard. He makes you stretch for a better and improved system.

Enjoy and learn from his teachings.

James A. Bickelhaupt
Director—Projects and Division ISO Coordinator
Mastercraft Fabrics
Division of Collins & Aikman Products Company
Spindale, North Carolina

Preface

As a representative of an independent third-party registrar, I cannot train, consult, or otherwise directly assist companies in system development or corrective action. To do so could present a possible conflict of interest should the company select us as its registrar at any time in the future. For a teacher at heart, this "arm's length" relationship with the student can be a bit frustrating. A few close friends, family members, and colleagues encouraged me to write about "ISO and all that registration stuff." It could be an outlet for teaching and may even be fun too, they said. But I procrastinated for four years. Though I've not yet reached my half-century mark (the big five-oh), I am, figuratively speaking, an old man in the ISO 9000 community, at least here in the United States. To play off a country song from a few years ago, "I was ISO when ISO wasn't cool." Allow me to share a brief walk through my vocational photo album.

In 1969, my career in the aerospace/defense industry began and continued until 1992, so those of you with like backgrounds can appreciate my affinity for the U.S. military standards. My employers were MIL-Q houses (MIL-Q-9858A). Virtually everything I have learned about quality has stemmed from that wonderful (now nearly obsolete) standard. I've been working in quality management since 1984. In 1989, my employer asked me to "find out what this ISO 9000 thing is all about," and in 1990 I took one of the very first certified lead assessor courses in the United States. It was so new, in fact, that the instructor was still based in the United Kingdom. Within the year I was certified by the Institute of Quality Assurance (IQA) in London, England. For you youngsters it is now known as the International Registry of Certified Assessors (IRCA). In its 1991 directory, I was one of only 50 or so credentialed pioneers in the United States. Of course, today there are hundreds.

During 1991 and 1992, I was privileged to do some ISO consulting and training and performed many system assessments as a subcontractor lead for a top ten registrar. I was growing to love the ISO 9000 standard as much as MIL-Q-9858. My credentials eventually included lead assessor certification with the Registrar Accreditation Board (RAB) and certified quality auditor (CQA) certification with ASQC. In late 1993, I engaged the ISO 9000 process in a full-time capacity, and in mid-1994, I relocated to the Charlotte, North Carolina, area to direct the registrar's interests in the Southeast United States. Presently, I'm the executive director, Southeast Region. I love the south. My snow shovel is now used to scoop up pine needles.

In 1994, the American automobile industry adopted the ISO 9000 series standards as the Big Three's QS-9000 requirements. This is proving to be a huge success for formal quality systems. I was fortunate to be in the first assessor training course offered by the auto sector group. It has been unofficially reported that the examination pass rate was very low for the first class—only one in five! I think it was only for God's goodness that I was in that top 20 percent. So, I became a certified assessor for QS-9000.

As the executive director for an international registrar, I am responsible for the Southeast United States. My challenge is to further the ISO 9000 registration process in the quality community, present credible information, promote my company's presence, and increase its business. Then there was my personal desire to teach and write my book. In late 1994, I coined "Meet the Registrar" as a solution. In general meetings (ASQC, APICS, Chamber of Commerce, at technical schools, and so on) open to representatives of many different companies, I could offer a broad range of material without conflict, entertain questions from the attendees and give them answers directly from the source (not second hand), and personally enjoy the teaching/presentation aspects. Meetings would last for one to three hours. During the first half, I'd give a canned presentation, then field any and all questions for the second half. The icing on the cake was that the event would be a completely free service to the quality community. Travel and accommodation costs were covered as marketing expenses. I provided the time and expertise, and sacrificed my 32-inch waistline to too many restaurant meals.

There was just one small glitch. What material was I going to present? I could not use someone else's. Whoever said "necessity is the mother of invention" had me in mind. Over the next few months I spent many an evening, and dozens of airline flights, developing chapter outlines for my book. These would become the essence of my presentations.

Meet the Registrar sessions became, and still are today, a great success. I've enjoyed sharing with many hundreds of people, and it has been my privilege to speak at some great seminars. This book is, in part, a collection of my works and material, a compilation of my presentations, from those many engagements. While my views and opinions do not necessarily represent those officially held by Bureau Veritas Quality International (BVQI) (NA), Inc., they do however, reflect the knowledge I've gained from more than 100 assessments and through my position as executive director. I hope you'll find them valuable as you prepare your quality system for ISO 9000 or QS-9000 registration.

Acknowledgments

Special thanks to a great friend, David Varella, for pressing me for a commitment to write this book. Thanks to my children and to Richard Lawson (DSM Chemicals) and Danuta Highet (BVQI (NA)) for added encouragement. Thank you, Jim Bickelhaupt (Mastercraft Fabrics), for writing the foreword. I'm grateful to the many fine associates at BVQI (NA); I've learned much from them. Thank you to Gary McMillan and Saft America for permitting me to reproduce your management review summary report. I appreciate the nice compliments and constructive criticism by the manuscript reviewers through ASQC Quality Press. Thanks to the many BVQI (NA) clients whose quality systems I've had the opportunity to assess. Above all, thanks to my wife, Adrienne, for her support and patience through the many nights and weekends of writing.

Note About the Standards

Throughout this book, I make mention or refer to the ISO 9000 series standards and the QS-9000 requirements. Rather than rely on footnotes on each page, I have chosen to cite them here. The standards are

ISO 9001, *Quality systems—Model for quality assurance in design, development, production, installation and servicing* (Geneva, Switzerland: International Organization for Standardization, 1994). The U.S. version may be obtained from ASQC.

QS-9000: Quality System Requirements, 2d ed. (Detroit, Mich.: Chrysler Corp., Ford Motor Co., General Motors Corp., February 1995).

ISO 10013, Guidelines for developing quality manuals (Geneva, Switzerland: International Organization for Standardization, 1995).

Introduction

What's the matter? All those boring reference books got you down? Fed up with the stuffy jargon? Looking for some commonsense, practical examples straight from the source? Not sure your expensive guru is right on? Looking for a second opinion before putting your quality system under the knife? You've been asking for a great chocolate cone and getting only pistachio macadamia nut—yuk! Well cheer up, friend, and Meet the Registrar!

This book was written out of a sincere desire to teach people about quality systems and about how to achieve registration for their companies. While there is already considerable information about ISO 9000 and QS-9000 available to the public, most of it is from secondary sources. Let me explain. Information currently in press or being presented at ASQC meetings, at seminars, at community colleges, at Chamber of Commerce gatherings, and the like, has *some* value. But, the source is usually a consultant, educator, internal quality auditor, or an employee of a registered company—all of whom are somewhat removed from the heart of system registration. Until Meet the Registrar, virtually nothing has presented the registrar's perspective. This collection of presentations is different. It offers a unique perspective through the eyes of a third-party registrar's executive director and senior lead assessor.

After all, it is the registrar, and no other source, that evaluates and ultimately passes judgment on a company's quality system and subsequent certification. You and I know that the final determinations about a company's quality system, and its eventual registration, will ultimately be made by a registrar, and not by any of those sources just mentioned. I believe than a responsible registrar should make the quality community knowledgeable about the registrar's own perspective on all things ISO and avail itself to answer questions from the community.

And what's wrong with a little humor? How about some examples from everyday life to which we can all relate? You'll find this book is not the same old, same old. But I promise you, your level of understanding will take a big leap. If not, the next time you see me, I'll buy coffee.

1

Management Representative: Authority Defined

ISO 9001 paragraph 4.1.2.3 states the following:

> The supplier's management . . . shall appoint a member . . . who irrespective of other responsibilities, shall have defined authority for:
> a)   ensuring the quality system is established, maintained . . . and,
> b)   reporting on the performance of the system . . .

Webster's dictionary says of the word *define*, "To determine the essential qualities or precise meaning of; to discover and set forth the meaning of."

This definition certainly suggests that the ISO management representative's authority should be described in more detail than just a simple *declaration* that he or she has authority and responsibility. And, there should be more than a mere replay of the wording in the standard.

This lead assessor looks for a well-documented enumeration of specific authorities, as well as responsibilities, that are unique to the ISO rep. These special responsibilities and authorities need to be clearly distinguished from those that are applicable to the individual's routine position or assignment. This distinction is especially true in organizations where the rep wears more than one hat and in companies where personnel are relatively mobile and transfers or re-assignments are likely. In these cases, the essential responsibilities and authorities of the rep can be readily associated with the position title, not with a particular person, and as such can be transferred to another individual with minimal confusion.

Some companies do well at defining and documenting a list of specific responsibilities, such as the following:

- Hold annual management reviews.

- Prepare and maintain the quality assurance manual.

- Ensure that ISO 9000 requirements are satisfied.

- Administer the internal audit program.

- Report on preventive action information.

- Ensure timely closure of corrective actions.

However, most companies usually neglect to balance these responsibilities with appropriate, commensurate authorities to permit, if not assure, that the person will be successful in carrying out his or her duties. Lack of clearly defined and documented authorities also brings about complications when a company finds it necessary to change reps.

Agreed, some authority is implicit in stated responsibility, but not always. And the implication is not always clear to, and interpreted the same by, everyone. Furthermore, while identifying the rep on the organizational chart certainly has value, this mostly just addresses the person's reporting relationship and may only serve to describe stature and prominence in the company. This does not ensure that his or her authority is defined, known, and understood by the rest of the organization. In the worst case, position on the organizational chart may only highlight and draw attention to a powerless, ineffective figurehead.

Webster's dictionary also defines *authority* as "The power to influence or command thought, opinion, or behavior." Some examples of explicitly defined authority that I have seen include the following responsibilities.

- Convene special sessions of the steering committee.

- Approve agendas for management reviews.

- Authorize corrective action implementations.
- Approve and assign internal auditors.
- Approve/reject interdepartmental system procedures.
- Make appeals to senior management.
- Critique customer complaint responses.
- Reassign quality personnel.
- Modify frequency and focus of internal audits.
- Change intervals and attendees of management reviews.
- Make corrective action assignments.
- Authorize the formation of project teams affecting quality.
- Recommend personnel changes.
- Determine ultimate disposition of nonconforming product.
- Assign/dismiss preventive action teams.
- Critique various departmental training programs.
- Initiate changes to all departmental quality system procedures.
- Authorize spending of specified amounts of money.
- Approve final acceptance of contracts.
- Approve company certificates of analysis/conformance.

Allow me to share three real-life examples where it was essential for the company to clearly define and document the rep's authority. Names and minor details have been changed for obvious reasons.

While personally leading a third-party quality systems assessment recently, the assessor and I were interviewing the quality assurance manager who also happened to be the ISO 9000 management representative for this company. It is often, but not always, the case that a facility's senior quality officer will be the management rep. We discussed his two respective roles and quickly concluded that he could not distinguish the two. It was not clear to him, to the assessment team, or to the rest of the company's management team, what his actual role was, or should be, as the management representative (not as the quality manager). Beyond the realm of his own quality department, this gentleman obviously did not have the appropriate authority or "the power to influence or command thought, opinion, or behavior" throughout the organization. He was, in effect, only a quality manager, and not a genuine management representative as intended by the ISO 9000 standards.

I'm very pleased to say that the company took the matter under serious consideration and remedied the deficiency before its eventual certification. It successfully determined and documented in a procedure those special responsibilities and authorities that the management rep needed to equip him to effectively implement and maintain the quality system and to report on its positive performance.

In the second example, the issue was related to the assignment of a new management representative, replacing an individual with long standing in the company.

Giant Gadget Company had a formal quality system that was created and implemented by its beloved management rep, "Mr. X." He then successfully led the first of Giant's several facilities through ISO 9000 pre-assessment and registration audits and gained ISO 9000 certification. But now that Giant's quality system was registered and preparing to enter the routine maintenance mode, Mr. X felt comfortable proceeding with his retirement plans. The company assigned him to a transitional role and appointed a replacement management representative, "Ms. Y." During the first surveillance audit, the important role of the management rep was evaluated. The assessors found that Giant Gadget had really experienced much more than a routine changing of the guard or passing of the baton.

While Mr. X was around, he actually had other duties associated with engineering, quality, and manufacturing that were combined with his role as the management rep. On the other hand, Ms. Y held a traditional total quality management (TQM) role, with heavy emphasis on training and human resources. The two very diverse roles of Mr. X and Ms. Y made it especially important to distinguish the daily assignments from the special responsibilities of the management rep—"irrespective of other responsibilities."

The level of effectiveness, authority (real and perceived), communication, and interaction previously enjoyed by Mr. X (primarily because of his technical orientation) was not automatically bequeathed to Ms. Y. Hers was a different world. Her background, network communication channels, and product and process knowledge were unlike (though not inferior to) those of Mr. X. Furthermore, Mr. X had been with Giant Gadget for all of his career. He knew the system and processes so thoroughly that he could cut through organizational boundaries and red tape with the greatest of ease. The quality system was his to command.

Ms. Y, however, had been with Giant for only a fraction of that time. Her company political alliances (yes, they are real) were not as well established. She had not the rapport with those in the technical side of the business that her predecessor did. She knew far less about the quality system and its various aspects and disciplines than did Mr. X. While she was very competent in her own right and in her own domain, she was, nonetheless, very different from Mr. X—and everyone knew it. As a result, her authority was also perceived as being different. To a large extent, this problem was because Giant had failed to clearly define and document specific authorities and responsibilities for its management representative.

At the conclusion of the assessment, Ms. Y expressed pleasure in knowing that finally her position authority would be understood by everyone. Ms. Y has since successfully led other Giant Gadget facilities to ISO 9000 certification, and its quality system is being well maintained under her charge.

In the last example, the concern was not related to the assignment of a new management representative, but rather to a system integration problem and to a duplication and overlap of roles.

The Big Balloon Company, prior to ISO 9000 implementation, had a well-established quality initiative—a TQM program managed by the TQM chairman. When its formal quality assurance system was being implemented, there arose some system integration problems between TQM and ISO, further complicated by the assignment of an ISO management representative different than the TQM chairman. Their respective individual roles and their associated authorities and responsibilities were unclear not only to themselves, but also to others at Big Balloon.

It was not until Big Balloon's pre-assessment audit that the issue was properly addressed. During early discussion with management, the requirement in paragraph 4.1.2.3 of the standard was made clear. By the closing meeting, it was apparent that many of the system integration problems could be linked, at least indirectly, to poorly defined and inadequately documented position descriptions. The managers needed more detail and specificity to allow the management rep to operate effectively and in harmony with the company's TQM chairman.

Big Balloon took corrective action that gave both people clear roles and responsibility, along with necessary authority. This was reflected in its system documentation. Big Balloon has since completed its formal registration audit, was certified, and now continues to maintain a healthy, effective quality system.

In conclusion, consider the following scenario, which happens every day in the best of companies. Through the eyes of a registrar, I frequently see similar situations.

Tomorrow, for whatever reason you care to imagine, your company will find it necessary to assign a different ISO 9000 management representative. Both likely replacement candidates are *not* from the quality organization.

Your company has always been very conscious about personnel and their training and development, so appropriately, the manager of human resources is a possible choice. She has an excellent rapport with the union, and she's very skilled at cross-functional team building. Though she is just beginning to learn about formal quality systems, she is well respected by most (but not all) of her manager peers. They think they are willing to support her.

The second candidate is the manufacturing manager, who previously has not been a pioneer for quality. But lately he's more supportive because of the progress he's seen the company make since its ISO 9000 registration. Also, he has been impressed by how the quality system, with its focus on process control and statistical techniques, has greatly improved the productivity of manufacturing. Mr. Manufacturing sure knows the operations side of the business. Unfortunately, he is often at odds with the engineering and purchasing managers. The director of research and development is very protective of his turf and has little use for anyone that tries to interfere with his department, especially Mr. Manufacturing.

Both nominees are viable choices. Both bring skills, strengths, and weaknesses. The final selection will be made by a simple vote from all members of management. The staff members have asked that, before they vote, that they know exactly what will be expected of the person in the new position. Even more than the responsibilities, they are interested in what power this person will have. They want to support the quality system, but they really don't want to be told what to do by anyone but themselves.

Your challenge, as the company president, is to identify the unique duties for which the new management representative will be responsible. More important, you have to determine what authority(s) this position will have. And, as one last bit of bad news, there is a hiring freeze. That means that whichever candidate is selected, he or she will have to assume the new role while maintaining his or her old one.

The responsibilities and authorities that you are about to define must be irrespective of the current ones associated with either manufacturing or human resources. And, they must be clear and well documented for the benefit of the voting staff, for the preservation of the quality system, and for the success and survival of the new management representative.

2

Assessment Readiness: What Do Registrars Look For?

If you think that your company's quality management system is ready for an assessment by your independent third-party registrar, you should be prepared to answer for yourself the question, "What do registrars look for?"

There is certainly no short answer to this question. But to generalize, the registrar is looking for everything that the assessment team sees, hears, reads, senses, feels, or otherwise observes during the conduct of an audit. To be a little more specific, I have dissected the whole elephant into seven bite-sized categories: documentation, implementation, compliance, effectiveness, efficiency, maturation and improvement, and evidence. Expressed as questions, they might look like the following:

1. Has the quality system been adequately documented?

2. Has the quality system been fully implemented?

3. Are the system's activities and practices compliant?

4. Is the implemented quality system effective?

5. Is the implemented quality system efficient?

6. Is the quality system maturing and improving?

7. Is there sufficient objective evidence of items 1 to 6?

It is essential to remember that any quality system assessment is only a sampling of observations. And be aware that this probably will not be a statistically valid sample. So don't be alarmed or openly defensive if you think the assessors may be overlooking some of the many excellent aspects of your system. The sample will be adequate to permit the lead assessor to be comfortable with the conclusions that he or she reaches. A larger sample may be taken if necessary in order to arrive at fair, unbiased, and accurate findings. The registrar's assessors are, like you, professional people. It is on these observations and resultant conclusions that registrars will base the final decision of whether or not the quality management system warrants ISO 9000 (or QS-9000) registration/certification.

In this chapter, we will further analyze each of the seven assessment categories to better understand what the registrar is expecting to see. It is nearly certain that, during the audit, these seven will not be addressed independently or in sequential order. Because we are talking about a quality *system*, documentation, activities, records, and results will be well mixed, integrated, and contiguous—as they should be. Consequently, all seven will have to be evaluated together, throughout the assessment. This makes the assessor's task more difficult, but also more interesting. This approach will also yield a clearer, more meaningful evaluation. Let's begin with documentation.

1. Has the Quality System Been Adequately Documented?

Quality management systems must be documented, formal (not bureaucratic), operating processes. If not adequately documented, they will never remain consistent. Without consistency, control and improvement are unlikely. Naturally then, the ISO 9000 standards require that the system be documented. What sort of documentation are we talking about? Usually there are four levels, or tiers, of documentation. While I sometimes see systems described with only three levels, I've never seen more than a five-tier approach.

The standard is not prescriptive here, but it does discuss at least three: the quality manual, quality system procedures, and quality work instructions. The most common document found in level four is quality system forms.

The Quality Manual

The quality manual is the highest level of all quality documents. It is sometimes referred to as the *quality policy manual* or the *quality control manual*. The name is not that important. The fundamental purpose of this high-level document is to

- Describe the quality policy or mission statement.
- Describe the scope of the system—what is covered.
- Define the structure or architecture of the system.
- Personalize or tailor the system to an individual company.
- Reflect the spirit or essence of the standard.
- Address all elements in the standard.
- Reference/include system-level procedures.
- Present commitments of the company.

Don't be tempted to adopt some other company's quality manual or to purchase a generic quality manual through your favorite quality magazine. Simply changing the title and headings will not suffice; the registrar will see right through that tactic. The quality manual needs to really be *your* quality manual, reflective of your company and describing your system. ISO 10013 is an excellent guideline for the format and content of your manual and procedures. The registrar will expect to obtain an official, controlled, and approved copy of the manual before the formal assessment begins. Oftentimes, an off-site adequacy documentation audit is performed on the manual in order to expose any errors or omissions. This early warning affords you courtesy and the time to make necessary changes before they manifest themselves as nonconformances. Also, advance notice will permit you to cascade the changes down into the system-level procedures as appropriate. Once the review is complete, most registrars have the lead assessor stamp and date each page of the manual. This effectively freezes or takes a snapshot of how the manual looks at the time of the formal assessment.

Future changes to the quality manual are usually handled in one of two ways depending on the magnitude and significance of the revision. The practice may vary with the registrar. For minor issues, the registrar may elect to review and stamp the new or revised pages at the next routine visit. If so, be certain to keep the old stamped sheets for reference. Or, the registrar may ask for official controlled updates through the mail. They will be returned after review and stamping.

If there are significant changes to the system, which definitely need to be reflected in the quality manual, the registrar will probably want to know right away—not necessarily for approval, but for awareness. Such dramatic changes might affect your registration or be cause for a special visit. It's better to take the high road and inform the registrar if you are in doubt. The stamped pages also serve as a flag if something is changed and the registrar has not been informed.

Regardless of how the revision is processed by the registrar, don't forget that the quality manual must be a controlled document. As such, all changes must be approved by your authorized people and there needs to be a well-documented revision history, describing the nature of changes (see the document and data control clause in the standard) for the registrar's review.

Another requirement of the standard is that the quality manual describe the *structure* of the system's documentation. *Architecture* or *design* are other terms that come to mind. Your registrar is not nearly as well acquainted with your quality system as you are. You may be surprised to find that some of your company's employees don't fully understand the system's documentation structure either. Consequently, it needs to be clearly defined within the quality manual. What sort of things should be considered?

Here are some questions the registrar may ask.

- How many levels of documentation are there?

- What is the architecture of the various levels?

- What type of system documentation is included within each level?

- Are forms an integral part of procedures and work instructions?

- Are there stand-alone forms?

- Do departmental-level procedures support system procedures?

- Is there direct correlation between the elements in the standard and the manual and procedures?

- Has the system provided a cross-reference between elements in the standard and procedures?

- What naming or numbering convention allows association between the standard and your quality manual, procedures, work instructions, and forms?

- Do separate operational units or departments have their own system procedure manuals, work instructions, and so on?

- Is documentation ordered along element numbers in the standard or along your departmental, product, or process lines?

- Are there any special-purpose addendums or supplements to your quality manual that limit or expand the basic system scope?

This is not an exhaustive, all-inclusive list of questions, but it should give you a fairly good idea of what everyone, especially your registrar, should know and understand about your quality system's structure. Your registrar will have to come to that understanding very early in the assessment process. The initial off-site review of the manual will be a starting point, but complete understanding will probably not come until the registrar has seen some examples of the other documentation.

Quality System Procedures

Quality system procedures are the most common type of second-level documentation. They too must be official, approved, and controlled documents within the quality system. These procedures give direct support to the quality manual. More important, they must be subordinate to everything prescribed and declared in the quality manual. Very rarely should they make a special allowance or provision that has not already been given credence in the quality manual, either directly by declaration, indirectly by inference, or by intentional omission or deference to the procedures.

The primary purposes for the quality system procedures are to

- Provide an overview of system-level activities and practices.
- Describe the who, what, why, where, and when.
- Define key responsibilities.
- Present systemwide definition and description.
- Identify lower-level supporting work instructions.

System procedures

- May be formatted or ordered by element of the standard
- Are sometimes arranged by department or business unit
- Need to cross-reference other pertinent/applicable procedures
- Need to link to, or be included/referenced, in the quality manual

The ISO 9000 series standards require that the quality manual either include or make reference to the system procedures. That is, you have an option here. Although I have visited a few companies whose quality manuals actually do include the system procedures, these quality systems are very rudimentary, are uncomplicated, and require only minimal documented definition. The manual can be sufficiently supported by only a dozen or two brief procedures. Too, these companies are relatively small in terms of both physical size and number of employees. For them it works quite well to have both documentation levels, the manual and procedures, combined into one document, or at least compiled in one common book. If the combination becomes so large as to be unwieldy and inconvenient, people are less likely to use the document, defeating its very purpose for being created in the first place. There is an added drawback to this approach that makes it undesirable. Whenever it becomes necessary to add a new system procedure or modify an existing one, the entire composite document is affected. The same, high-level review and approval of the quality manual must once again be secured. Granted, this is not usually a problem for a small company with a simple quality system.

You want your quality system to be convenient and flexible. It is appropriate that system procedures be revised from time to time. The quality manual, on the other hand, should be a fairly stable document, changing at a much less frequent rate than system procedures. For this reason, combining the two levels may not be a practical, effective option for your quality system.

If you elect not to actually *include* second-level system procedures within the quality manual, then you must *reference* them within the manual. The registrar will expect this reference to be much more than just making mention that there are system procedures, or simply stating "The manual is supported with procedures whose numbers align with sections of the manual," or "There are 20 system procedures, one for each of the elements in the standard," or "Our quality system is further described by procedures contained in our systems procedures manual."

Such declarations may be true and helpful, but they do not fulfill the intent of the standard. The reference requirement can be satisfied in a couple of ways that should be acceptable to your registrar. The most direct method of reference is to actually identify the procedure by name and/or number. It's a poor practice to specify the revision level of the procedure, however. If the procedure changes, then you are forced to change the quality manual as well. A second, perhaps more effective method of reference is to specifically call out, by name and/or number, an official, controlled document that serves as a table of contents or directory to all the system-level procedures. This type of document must be officially included within your quality system. As such, it must be created, reviewed, approved, and maintained by an authorized party. The table/directory, to be complete, must be a comprehensive listing of all system procedures and must provide their positive identification by either name and/or number and positively show the revision level of each procedure.

This reference method lets you simply update and approve the table/directory whenever a procedure is added, deleted, or revised. The quality manual can remain untouched. Again, don't specify the version level of the table/directory within the quality manual for the obvious reason already discussed. This method of reference provides a bridge, or official link, between the manual and the procedures. This method is acceptable, provided the table/directory is a legitimate component of the system documentation.

Quality Work Instructions

Work instructions are typically included in the third level of the quality system documentation. They support, and are subordinate to, system procedures. As such, they must comply with everything prescribed and declared in the procedures. Very rarely should they make a special allowance or provision that has not already been given credence in the quality procedures, either directly by declaration, indirectly by inference, or by intentional omission and deference to the work instructions.

The primary purposes for the quality work instructions are to

- Provide specific detail about activities and practices.
- Describe how things are to be done/accomplished.
- Describe the who, what, why, where, and when.
- Define key responsibilities.
- Identify other related pertinent work instructions.
- Identify lower-level supporting work instructions.

Work instructions

- May apply to administrative activities
- May be formatted or ordered by element of the standard
- Are sometimes arranged by department or business unit
- Need to cross-reference other pertinent/applicable instructions
- Need to link to higher-level, parent system procedures
- Immediately include, or reference, pertinent forms
- Are usually written, text-type documents
- May include or be flowcharts, diagrams, pictures, or models
- Are not limited only to production or machine operation activity

When determining the need for a work instruction, remember this rule of thumb: "If the absence of the work instruction may actually, or potentially, affect the quality and/or consistency of the work being performed, then the work instruction is probably necessary."

As with all other levels of quality system documentation, work instructions need to be official, approved, and controlled. However, the type and extent of control exercised over them may not be the same as, or as stringent as, that applied to system procedures and will probably not be as extreme as control over the quality manual.

When work instructions take the form of pictures, diagrams, models, specimen, and the like, their control will usually be different than that associated with traditional text documents. Regardless, their control must provide for approval by authorized persons, assurance that the appropriate version is used, and the other appropriate requirements for the control of documents and data as specified in the standard.

Quality System Forms

Quality system forms are generally contained in the fourth or lowest level of the quality system documentation, although they may be in a higher level when they are included as an integral part of a procedure or work instruction.

The primary purpose of quality forms is to

- Specify vital information recording.

- Capture essential data.

- Become quality records (usually, but not always).

Quality forms may

- Be stand-alone documents

- Be included in procedures and instructions

- Serve as self-explanatory instructions

When determining the need for a quality form, remember these basic guides.

- If the absence of the form may actually, or potentially, affect the quality and/or consistency of the work being done, then the form is probably necessary.

- If the absence of the form will prevent the capturing of essential information/data and prevent the creation of a required quality record, then the form is probably needed.

Remember though, not all records are quality records. Accordingly, not all forms have to be included as official documentation in the quality system. Refer to the section of the standard that deals with quality records. We'll further discuss quality records later in this chapter.

As with all other levels of quality system documentation, quality forms need to be official, approved, and controlled. However, the type and extent of control exercised over them may not be the same as, or as stringent as, that applied to system procedures and instructions, and will probably not be as extreme as control of the quality manual. Issues relating to the distribution, and possible recall, of forms may be treated differently than with other types of quality system documentation.

To bring our discussion on documentation to a close, think back to our question "Has the quality system been adequately documented?" To answer that for yourself, consider the following: Does the quality system documentation (especially its procedures, work instructions, and forms) have sufficient content and adequate detail? Commensurate with the following:

- Previous employee training

- Task difficulty and complexity

- Consequences of failure

- Established qualifications of personnel

To permit consistent and effective implementation by

- All affected personnel

- Present employees

- New employees

- Transferred employees

- Returning employees

Assuming that the registrar's representative is comfortable with your system's formal documentation, what is the next consideration? He or she will want to turn attention to implementation.

2. Has the Quality System Been Fully Implemented?

No doubt, it is a real accomplishment if you have succeeded in documenting your quality management system through a quality manual, procedures, work instructions, forms, and so on. It's quite another thing to actually *implement* the system.

An internationally known registrar, with whom I am familiar, conducted a study about two years ago in an effort to analyze the nature of nonconformances experienced by ISO 9001– and ISO 9002–registered companies. The data were collected from 26 companies. In total, there were about 300 nonconformances between all the companies. Among other insightful things, the study showed that about half of all deficiencies related to problems with documentation. The other half tied to implementation problems!

The registrar will expect that the implementation is of the same, identical quality system that is described in the system documentation. If, along the way, some other approach has been implemented, perhaps because someone has decided it's better than what is prescribed, then the implementation is in error. The registrar is not trying to inhibit creativity, but does want to ensure that it is legitimate and constructively channeled. If it is believed that the alternative is better, then the system should make provision for evaluating it. If it is determined to have merit and its benefit warrants making a change, then the official system can be revised to give the secondary method legitimacy. If not warranted, it can be discarded. Without this kind of structure, discipline, and validation, the system will erode into chaos. The system will cease to be a system. It will default to independent, inconsistent, fragmented activities.

Suitable implementation will mean that all applicable elements of the standard (as defined in the quality manual as being within the scope of the system) are active—fully functioning throughout the entire operation. Partial activity, or full activity but only in limited areas, demonstrates unsuitable, incomplete implementation. While this might be explainable in developing systems, it is nonetheless noncompliant. The only exception might be if the documentation defines the system in such a way that it makes allowances for more than one approach. But each different approach must still meet all requirements of the standard. Implementation must be consistent. All specified activities and practices must be performed the same way each time, unless otherwise specially permitted by the system. Inconsistency will eventually result in loss of control. An out-of-control system will be more difficult to improve.

Likening the system to a conventional manufacturing process, improvement is difficult, if not impossible, until the process is stable and controlled.

3. Are the System's Activities and Practices Compliant?

Paragraph 4.17 of the ISO 9000 series requirements standards specifies, "The supplier [like your company] shall establish and maintain documented procedures for planning and implementing internal quality audits to verify whether quality activities and related results comply with planned arrangements and to determine the effectiveness of the quality system."

Any system left to itself, quality or otherwise, will eventually deteriorate. At the very least, it will fail to conform in every aspect. It will exhibit signs of noncompliance in one or more areas. Even if the system is maintained and actually improves, there will always remain the possibility of isolated compliance problems.

Furthermore, the very best implementation plans are imperfect, resulting in probable inconsistencies and incompleteness. Apparently, the authors of the standard were of the same conviction; hence the requirement for internal quality auditing. Your registrar will require that your quality system demonstrate ample objective evidence of compliance as a prerequisite to certification/registration.

In paragraph 4.17, the standard refers to system compliance with *planned arrangements*. What might these planned arrangements be? Consider the following:

- System procedures

- Work instructions

- Prescriptive forms

- Quality plans

- Training materials

- Preferred practices

Is it possible that, under unusual or extenuating circumstances, the planned arrangements are not appropriate? Isn't it reasonable to design and implement a quality system whose routine activity accounts for only 90 percent of all possible situations?

The answer to both questions is yes; that is, unless an individual is inclined to

- Overspecify and overengineer.

- Create excessively complex systems.

- Disregard resource limitations.

- Prefer theory over practicality.

- Confuse burden with value.

Why would anyone intentionally create a quality system that attempts to plan for every remote potentiality, every contingency, regardless of its improbable likelihood of occurrence? Why indeed? Perhaps the consequences of failure are so severe, so extreme, so catastrophic that the greatest possible degree of control is justifiably demanded and unquestionably warranted. I can quickly think of a few such systems: those for nuclear facilities, aerospace hardware, medical procedures, health and safety equipment, pharmaceuticals, hazardous waste disposal, national defense, air traffic control, biological research, . . . You get the idea. Some of the most complex systems are truly justified (although even they may not ensure 100 percent reliability).

But such systems as these probably account for only a very small portion of all systems. Chances are great that your company does not require such a comprehensive, regulated quality system. For the majority of quality systems, it is acceptable to plan for only those conditions that are likely, or probable, to occur. That is, the standard controls and system disciplines that are routinely administered need only account for customary conditions—those that will probably happen, nearly every time. You may not even be able to think of all other possibilities, let alone intelligently plan for them. I am not suggesting that you ignore all but the most common and predictable situations.

Implementation must be *consistent*. All specified activities and practices must be performed the same way each time, unless otherwise specially stated. Those conditions that are unusual, but are nonetheless less possible with reasonable odds (not the rare) should at least be considered within your system. After some sort of cost versus benefit analysis, or risk/needs analysis, you should describe in procedures and work instructions how the exceptional cases will be handled. Describe such alternate things as authorities, plans of actions, documentation, approvals, equipment, process controls, sample sizes, inspections, sources, and materials. In those few cases where detailed written instructions may not be warranted, there needs to be clearly defined responsibility of affected people.

Adequate system definition will help ensure that unusual situations are appropriately controlled. Lack of documented description will allow order to default to chance, promoting inconsistency and forfeiting the compliance of practices and activities.

Simply put, when planning system activities and providing for their compliance, keep your head out of the clouds and don't stick it in the sand.

On those few occasions when you have to deviate from the traditional path and follow the alternate contingency plan, by all means make sure that the exception is duly authorized/approved and documented. This will be vital substantiation when the registrar's representative mysteriously selects, through his or her random sample, this one-in-a-hundred case.

So now that you have adequately documented plans for the common situations and possible (not improbable) occurrences, you have set the basic criteria against which compliance can be judged—not only by your own internal auditors, but also by your registrar. A registrar, and its professional assessors, will not be preoccupied with the personalities, appearance, or performance of your company's personnel. And as professionals, they are trained to make allowance for natural nervousness and human emotions. They are there to evaluate your quality system, not your employees.

However, the quality system is a system and, as such, must be implemented through people. The registrar will primarily assess compliance through interaction with, and observation of, personnel. But keep in mind, this will only reflect on the system and not on the people. Just how will the registrar determine whether the system's activities and practices are compliant? What will the assessors look for? Recall the beginning of the chapter. The assessors will acquire information through all of their natural senses. And the best auditors have a sixth sense as well. While remaining objective, assessors will seek to determine if

- Individuals are following and adhering to their procedures and work instructions.

- Personnel are correctly trained and equipped.

- People are aware of what they should be doing.

- Exceptions to prescribed methods/practices are authorized and documented.

- Alternate approaches are described in, and allowed by, procedures.

- Descriptions and explanations by interviewees agree with written definitions.

- Observed conditions align with procedures and instructions.

Consider this: Say what you do, and do what you say (provided the description and the practice is compliant with the standard, and is effective).

4. Is the Implemented Quality System Effective?

"The best laid plans of mice and men . . . " So goes the old adage. Sometimes, even the best designed systems, implemented by the most conscientious, skillful, and dedicated people, will yield results that are less than totally effective. One hundred percent perfection is rare, especially in human endeavors. Internal quality audits are one method prescribed by the ISO 9000 standards (see paragraph 4.17) for detecting ineffective system elements. Corrective action (paragraph 4.14.2) and preventive action (paragraph 4.14.3) are also required by the standard as a means to remedy ineffectiveness, for making the system more robust, for improving it, and for making it more effective. The degree to which the system will be effective depends on many factors, some of which are the following:

- Thoughtful and comprehensive system design

- Planning for probable and possible contingencies

- Thorough and consistent implementation

- Adequate resource allocation

- Skills and abilities of personnel

- Clearly defined and documented responsibilities

- Appropriate levels of delegated authority

- Employee training and development

- Compliance to system and standard requirements

- Sufficient time for system maturity and improvement

- Management commitment and involvement

Can a documented and implemented system meet the requirements of the standard and yet be ineffective? Not exactly. Personnel may comply with their procedures and work instructions and yet still produce ineffective results! Consequently, the registrar will look beyond mere compliance in order to ensure system efficacy. What are some of the telltale signs that a system is operating effectively? Well, the proof is in the pudding. Output or results of an effective system will be

- Desirable

- Predictable

- Expected

- Controlled

- Consistent

- Repeatable

Furthermore, the registrar will look to see if activities and practices meet the *intent* of the governing procedure and work instruction and whether or not they satisfy the "spirit" of the standard. As precise as the English language is, it is difficult for us to convey every thought with the written word. Your personnel need to understand more than the letter of the law. They need to have an appreciation for the intent of the requirement. Otherwise, effectiveness may be jeopardized.

Let me share a personal example that now, some five years later, seems a little humorous. My teenage daughter was leaving the house to go on her first real (without a chaperone) date. My last words to her were, "I love you. Be home at 10:00 P.M.!" "Okay, Dad." At 10:30 P.M. I heard her come in the back door.

"Hi, Dad"—without any sound of guilt, without hesitation.

"Why weren't you home at 10:00?"

"I was! We've been sitting in the car until now."

The intent (my intention) was that my daughter would be *in* the house, safe and accounted for, by 10:00 P.M. While she complied (technically) with my request, her actions did not effectively meet the intent of the rule.

Want another example a little closer to the quality system? Here's one.

I visited ACME (fictitious) Widget Company. The employees properly addressed in their procedure the standard's requirement for inspection and test status (paragraph 4.12). It requires some suitable indicator on inspected and tested product. The intent is that subsequent key operations (right through to shipping) know that the product has successfully completed the mandatory inspections and tests throughout its processing.

There it was, right in the work instruction. Only after the product passed voltage test would the inspector put the green sticky dot on the outside face of the subassembly. Great! The inspector followed the practice (complied) precisely. Green dot . . . green dot . . . green dot . . .

"Why do you do that," I asked.

"It's obvious, isn't it—to show that the subassembly passed the voltage test."

"Don't *you* already know that it passed, even without a green dot?"

"Sure I do. I did the test."

"So why the green sticky dot?" I probed. (I'm pressing my luck.)

"Oh, now I understand what you're getting at. So the final assembly people will know it passed. Final assembly is only done on third shift and I'm not here to tell them."

"That sure makes sense. Thanks!" Guess where I went. To third shift, later that same night.

"Hi, I see you are doing the final assembly and packaging of the subunits prepared on the first shift."

"That's right. I'm the last person to see it. After adding a few pieces of hardware, I give it a thorough visual inspection to make sure there are no missing parts, no dents, no paint scratches and such."

"What about the subassembly voltage test? Are you sure it was done? Are you sure it passed the test?"

"I don't know. That's done by the inspector on first shift."

"What about the green dot, don't you look for it?" I asked.

"Oh, *that* green dot! So that's what that thing is for. Yes, now that you mention it, I do recall seeing it sometimes, but I really haven't paid much attention to it. Anyhow, like I said, the first shift guy tests it."

"Suppose that in the future you noticed some without the green dot—would you hold them back or reject them?"

"H— no! I can't do that. I have to keep this line running. Besides, only the tester rejects it if it's bad."

What happened here? The system prescribed the application of a very suitable inspection and test status indicator. The first shift inspector complied. But was the system control effective? Definitely not! The intent was missed all together. Why? Here are a few reasons.

- Deficient final assembly/visual work instruction

- Insufficient operator training

- Lack of understanding of intent

- Unclear responsibility

- Unclear authority

If the person who should be looking for the presence (or absence) of the green dot doesn't know to do so, he or she will never detect a subunit that either failed or skipped the voltage test. Further, then, there is absolutely no value in applying the dot. Detecting missing dots would be of no use, as the final visualler didn't know he or she had authority to take action. The entire back half of the system control method was totally ineffective. It's the weakest-link-in-the-chain story again.

Another characteristic of quality system effectiveness is objective evidence that the system meets the "spirit" of the appropriate ISO 9000 standard. Your registrar must look for this evidence. The standard is much more than a compilation of rules and requirements. It provides a generic model for a quality system. Its spirit is communicated throughout the document, but especially so in the scope statement, paragraph 1. Paraphrasing, it says that the standard is for use where a supplier's capability to design and supply conforming product needs to be demonstrated, and the requirements are aimed primarily at achieving customer satisfaction by preventing nonconformity at all stages from design through servicing. Minimum compliance to the written letter in the standard will not necessarily satisfy the spirit of the document. Overall effectiveness of your quality system will be judged, in part, on how well it emulates the standard's spirit as expressed not only in the scope statement, but also in its 20 elements.

Many registrars have prepared their own interpretive aid that defines what they believe to be the spirit of each of the elements. Such a document may be called something like *sanctioned interpretations*, *prime intents*, *core themes*, or *fundamental aims*. Though these documents are rarely ever published except for internal use, your registrar may be willing to discuss them with you. Don't forget, it is the registrar that will make the final determinations and judgments about your system. You should become acquainted early on with your registrar's perspective.

Probably the most tangible, objective evidence that your system is operating effectively is *quality records*. Of such records the standard (paragraph 4.16) says "[they] . . . shall be maintained to demonstrate conformance to specified requirements and the effective operation of the quality system." Virtually every activity prescribed by the system must be substantiated with documented records. But simply having quality records, or even an abundance of them, does not in itself mean system efficacy. Effectiveness, or ineffectiveness, will be proven out by the content of those records. They will either support or refute how well various aspects of the system's activities are being performed and how effective their controls are. Records also reflect on the efficacy of the overall system. For instance, consider the following:

- Records of design reviews and validations reflect on design planning and execution effectiveness.

- Records of receiving inspection reflect on the effectiveness of subcontractor selection and control.

- Records of final inspection rejections reflect on in-process inspection and process control effectiveness.

- Records of corrective action reflect on system implementation effectiveness.

- Records of customer complaints reflect on overall quality system effectiveness.

- Records of internal audits reflect on overall quality system effectiveness.

- Records of defective product returns reflect on the effectiveness of nonconforming material control.

Registrars will definitely want to look at your quality records. They will want to see enough records to show ongoing and sustained system operation. The registrar's assessors will be interested in what your own records say about your system's effectiveness. Where data are obtained and documented in quality records you, the company, can help define what is effective by identifying the desired outcome (expected results, performance limits, and so on) and then comparing this against actual results obtained. Quality records are the assessor's best friend.

Finally, the results of your *internal system audits* (and process and product audits, if you do them) will tell a great deal about system efficacy (not only compliance). If the audits are meaningful and properly performed, they will not only identify weaknesses but also strengths. As a lead assessor, I have a particular method of operation or style/approach when it comes to assessing internal audit activity. I almost always schedule this element very near the end of the audit. By doing so, I can never be unduly influenced by any of the company's own findings (good or bad) described in audit reports. Too, the auditee will know I'm not influenced. I think this approach gives added objectivity to my assessment observations and nonconformances and will probably add validity and credence to your audit program and your auditors, especially when they have findings similar to mine. My approach is not the only correct one. Some excellent assessors choose to look at internal audit records very early in the audit. The assessor will then use this information as input for further evaluating the effectiveness of the audit program, of the corrective action system, and of the management review. Regardless of the lead assessor's style, he or she will learn a lot from your audit records. They reflect on the overall compliance of the system, on its effectiveness, and on the effectiveness of its component activities.

In summarizing system effectiveness, remember that any job worth doing is worth doing well! Ask the following questions.

- Is the job getting done?

- Is the job getting done well?

5. Is the Implemented Quality System Efficient ?

We've just finished talking about system effectiveness and what the registrar looks for. So now, what's the concern about efficiency? Well, first of all, effectiveness and efficiency are two related but different attributes. Efficacy does not automatically mean efficiency. Your system can yield effective results, but at what expense? If the cost is too great, system activities may not be efficient. The extent to which efficiency is a concern will depend on the particular registrar. Some may assess your system only for compliance and effectiveness and have little or no interest in how efficient it is. I believe that the best registrars will also look at how efficiently your system is functioning. The registrar with whom I am acquainted has an unwritten motto of "value, not burden." The registrar can bring added value to your operation by exposing not only ineffective, but also inefficient, methods and practices pertaining to your quality system. Think about it. Your registrar's assessors have probably each seen more than 100 quality systems and remember the most outstanding ones. Alhough they should never impose another's system on you, why shouldn't you take advantage of their knowledge?

Will you get a nonconformance if one or more components of your system is not operating efficiently? Probably not, as long as it remains compliant and effective. But your registrar is not giving you full service if the auditor ignores the inefficient condition. At the very least, the assessor should raise it as an observation, making mention of it in the closing meeting and/or in the final audit report. And you are cheating yourself if you don't take the comments to heart. I also have the opinion that the best registrars will call attention to the good, positive things being done; this will reinforce effective and efficient practices.

What will the registrar look for to determine efficiency of the overall quality system and of its component functions? What is system efficiency anyway? Let's begin with a definition. *Efficiency* is the realization or attainment of *effective* results with only a reasonable amount of resources. So while system activities can be effective without being efficient, in order to be efficient, they must also be effective. The registrar will consider just what amount of resources is being consumed to produce the effective results. Obviously, the fewer resources, the greater the efficiency, generally (so long as efficacy is not compromised). I am talking about all types of resources, including people, materials, equipment, energy, time, skill levels, and technology.

Another point of interest will be redundancy. The most efficient systems and processes will have identified and reduced or eliminated redundant activities. Purpose and meaning also reflect on efficiency. Again, the best systems and processes have identified and reduced or eliminated activities that are not purposeful and meaningful. The registrar will endeavor to point out these unnecessary functions, tasks, and so on.

As mentioned before, your quality system is implemented through people. And, for most companies, their people are a valuable resource. But that value may not be fully realized even though the potential exists. The registrar must determine if personnel whose jobs affect quality (not just the quality department, not just product or service quality, but the company's entire quality initiative and quality system—Big Q) have been adequately trained to meet the minimum requirements of their position. However, I have observed that the most efficient companies and quality systems are those that have further trained their employees to maximize the use of their latent talents and to realize their fullest potential.

Efficiency is contrary to waste. Waste is expending unnecessary resources. Unnecessary may mean too much, the wrong kind, or the wrong level. The registrar, if concerned about efficiency, will look to see if the appropriate amount, type, and levels of technology, equipment, materials, and personnel have been utilized. If inappropriate or unnecessary resources are consumed, the results may still be very effective. But because of the waste, the results are not efficient. Overkill, overutilization, or overspecification promotes inefficiency.

For example, the system generally should not be using an engineer to sort out parts for rework or an electronic technician to repair inspection lamps, or a marketing director to enter orders at a computer terminal. While they may be capable to do these tasks, these employees are greatly overqualified. Using them would be wasteful, except in unusual circumstances. Using a laser ruler (accurate to millionths) when a micrometer would suffice, or an electron microscope when a magnifying glass is fine, is applying equipment in an inefficient manner. Using a biologist to analyze metallurgical problems, or a statistician to address a material handling problem, is probably not the best choice of personnel type; again, it is wasteful. Specifying tolerances to four decimal places when ±0.005 is adequate, or requiring a C_{pk} of 2.0 for an insignificant characteristic, is unnecessary overspecification. Using titanium when aluminum is fully adequate, or deionized water in a water-based machine coolant system, is likely to be a waste of materials. I'm sure these simplistic examples are obvious to you.

But look closely at your system, and you will surely find more subtle—but equally inefficient—cases. Look for appropriate types, levels, and quantities of technology, equipment, materials, specifications, and people—appropriate for the need. That will tell you, and the registrar, about your quality system's efficiency.

One last thought on efficiency. The thought for today is justification. Do the results justify the means? Does the output warrant the input? Does the reward exceed the cost? Test every aspect of your company's quality system, and its various activities, against these three questions. Remember: value, not burden! Inefficiency means lost revenue.

Consider this example. The task is to nail 100 boards together, in sets of two. Here are three possible solutions.

1. Use a sledgehammer and two people. This is effective, but definitely not efficient.

2. Use a claw hammer and one person. This is somewhat effective and efficient.

3. Use a pneumatic nailer, and one person for 1/10 the time. This is the most effective and efficient solution.

6. Is the Quality System Maturing and Improving?

As we've discussed so far in this chapter, a registrar will be looking for an adequately documented quality system that has been fully and consistently implemented, that is being complied with, and that is producing effective results. That's a lot! And you'd think any registrar would be more than satisfied to find such a system. Not necessarily. Much of the strength of the ISO/QS certification and registration scheme relies on the credibility, independence, and recognition of the third-party registrar. To a large degree, all three are realized when the registrar is associated with, and accountable to, a higher authority. The highest authority in the scheme is the national accreditation body, which almost always is an affiliate of its country's government.

A directly accredited registrar (one that enjoys a firsthand relationship with, and can issue certificates on behalf of, national accreditation bodies) will look to see that, over time, the quality system is not only *maturing* but is also *improving*. The accreditation bodies require the registrar to do so. Continued, minimal compliance is not acceptable. Conditions that are marginally adequate and allow for the initial registration of the quality system may not be approved during subsequent maintenance visits by the registrar. Many of the elements in the standard, even if fully met, will only permit the quality system to remain status quo. The titles of these elements often include *control*, such as the following:

- Control of nonconforming product
- Control of documents and data
- Control of inspection, measuring, and test equipment
- Control of customer-supplied product
- Process control
- Design control

Such elements are regulatory and compliance oriented. Then there's another group of elements that, when fully embraced, can take your system beyond where it is today. The QS-9000 requirements are particularly concerned with forward progress. QS-9000 devotes a whole section to continuous improvement, requiring the philosophy to be deployed throughout the entire organization. This latter group is proactive, motivational, and improvement oriented. Such activities include the following:

- Corrective action
- Preventive action
- Quality planning
- Management review
- Internal quality audits (if done with proper motives)
- Oversight by the management representative
- Setting quality objectives
- Training (if developmental in nature)

- Continuous improvement

- Process audits (operational and administrative)

How well this last group is exercised will determine the degree to which your quality system will mature and improve. Merely sustaining the system over time does not imply that it is any better than it was on day one. If the system is truly moving along, accelerating, and gaining momentum; if it is enabling the business to become more efficient; if it is delighting customers, reducing waste and increasing profits; and if it is experiencing the many benefits of success, then there will be ample evidence for the registrar. Let's discuss some in detail. The registrar will look to see if the management reviews are

- Regularly conducted

- Well attended

- Supported by executive-level management

- Exhibiting senior management participation

- Identifying needs/opportunities for system change

- Meaningful and insightful

- Insisting on correction to problems

- Promoting prevention and robustness

- Properly documented

Are all appropriate sources of quality data and records being collected, reviewed, and analyzed for preventive/proactive opportunities? Are the resultant actions and information reported to management?

Are system deficiencies and individual cases of nonconformance being addressed through appropriate levels of formal, documented, corrective action commensurate with the problem's degree of criticality and risk?

Does the documentation substantiate that the corrective measure was actually taken and that it was effective in resolving the problem and in avoiding its recurrence?

Do all impacted personnel, and those whose jobs affect quality, have the opportunity to give constructive input/criticism? Do they have the means, or is there a formal scheme, for initiating changes to the system? Are their improvement ideas accepted and implemented when warranted?

Are the quality system audits identifying genuine issues? Are they exposing the few significant systemic problems and raising real concerns or merely just rehashing the many trivial nits? Are they regularly and aggressively exercising all elements of the system? Do they seek and find improvement opportunities, or just focus on minimum compliance? Remember: value, not burden!

Is progress being made toward the established quality objectives? Are improvements being seen in the measurements of process and product quality? Are efficiency and productivity levels improving?

Are "off quality" indicators such as the following going down?

- Scrap, rework, and repair
- Customer concession requests
- Returned nonconforming goods
- Customer complaints
- Late deliveries
- Setup errors
- Design flaws
- Lot inspection failures
- Purchased goods rejects
- Equipment downtime
- Inventory cost of work in process (WIP)
- Internal material review board requests

Are "on quality" indicators such as the following going up?

- Process capability indices (C_{pk})
- Yield rates
- First-time-through ratios
- Vendor performance ratings
- Productivity factors

- Product audit pass rates

- Customer satisfaction ratings

Are administrative processes and tasks being improved and streamlined?

- Reduced nonmanufacturing lead times

- More accurate contract reviews

- Fewer planning/design errors

- Shorter specification review cycles

- Less time to process quotations

- Simplified vendor selection and approval

- More timely complaint resolution

- Quicker processing of engineering changes

- Fewer order entry and shipping errors

- More accurate/thorough purchasing documents

- Shorter process planning periods

Are personnel skills being improved and developed?

Is quality system documentation being regularly reviewed and updated as appropriate?

It is easy to discern dozens of indicators that reflect on quality system maturity and improvement. Virtually every indicator will be validated by your quality records, internal documentation, and associated data. The registrar's assessors will want to look at a sample of them.

7. Is There Sufficient Objective Evidence of Items 1 to 6?

Recall what was said at the beginning of this chapter. The registrar is looking for everything that the assessment team sees, hears, reads, senses, feels, or otherwise observes during the conduct of an audit. The assessment team is collecting objective evidence throughout the entire visit. Actually, the evidence collection begins even earlier, with the off-site review of your quality manual. It continues right through to the formal closing meeting.

The disposition of the auditors, the temperament of the audit, and how the evidence collection process unfolds will depend predominately on the mind-set or primary mission of the registrar. Let me clarify. Registrars that emphasize compliance auditing and are predisposed to finding nonconformities may be bent on identifying what's wrong with the system, rather than what is right. They may be inclined to seek out every minor or insignificant infraction to justify why they think your system is not properly functioning. That's why there are horror stories about registrars that write 100 to 150 or more nonconformances. (By the way, some of those horror stories are true!) Fortunately, while these registrars do exist, they are in the minority.

The better, more sane approach is to look at the evidence with an eye on conformance: discerning how the system can, and is, satisfying the standard. If there are system deficiencies, then write system nonconformities—not one for each and every occurrence of the same or related deficiency. Once again, "value, not burden." Registrars in the preferred category will typically write 10 to 30 or so nonconformities, but this number can vary. I'm familiar with cases that ranged between zero and 38, but this is still a far cry from 100 to 150.

The registrar must be able to support the *audit findings* with an ample collection of objective evidence. Notice I didn't say *audit nonconformances*. Yes, nonconformances need to be substantiated with objective evidence of noncompliance or deficiency. But positive results also need to be supported. Compliance and effective/efficient activities should be demonstrated with objective evidence. Without irrefutable, objective evidence to support conclusions, the assessor may just be guessing, or may only be responding to last night's spicy Mexican dinner. Then again, he or she may just like you or not like you!

There are really only five types of objective evidence that the audit team mambers can acquire through their natural senses. They are the following:

1. Observed practices

2. Observed conditions

3. Personal discussions

4. Personal testimonies

5. Documentation (quality records)

To recap and summarize this chapter, when performing an independent, third-party, quality systems assessment, the registrar will look to see if the system is

- Documented
- Implemented
- Compliant
- Effective
- Efficient
- Maturing
- Improving
- Evident

3

Setting the Scope of Your Training Initiative

ISO 9001 paragraph 4.18 states the following:

> The supplier shall establish and maintain documented procedures for identifying training needs and provide for the training of all personnel performing activities affecting quality. Personnel performing specific assigned tasks shall be qualified on the basis of appropriate education, training, and/or experience, as required. Appropriate records of training shall be maintained.

That's basically the extent to which the standard addresses training, except for a very brief mention in a couple of other elements. This brevity must in no way be misconstrued to mean that training is any less important in the minds of the standard's authors than the lengthier elements. Training will certainly be of equal interest to your registrar.

The essence of the requirement is that you identify, plan for, and then execute the training activities of pertinent personnel *whose jobs affect quality*. Don't underestimate the scope of the last part of that statement. It does not limit the scope of applicability to only those people in the quality department, or to only those people that interface with, or otherwise impact, the quality organization. In fact, the focus is not on the quality department or organization as an entity. The standard, when it refers to quality, is talking about the Big Q, if you will. It's talking about quality as a discipline throughout the entire company. Don't limit your thinking to just product quality or service quality either. Consider all processes and activities, both administrative and operational, that are directly prescribed by the standard or by your company's own quality system. Also consider those processes and activities that are indirectly influenced or affected by those that are prescribed. Now you are getting a better appreciation for the breadth and depth of training's scope.

Let's consider a few examples of the kinds of positions and individuals that do apply. The following list, though far from being complete, includes some of those probably affecting quality.

- Production operators
- Inspectors
- Purchasing agents
- Quality management
- Manufacturing
- Internal auditors
- Design engineers
- Calibration technicians
- Equipment preventive maintenance techs
- Packers/shippers
- Material handlers
- Service representatives
- Machinists
- Manufacturing supervisors
- Quality engineers
- Field engineers
- Quality planners
- Marketers

- Draftspeople
- Order processors
- Process engineers
- Trainers
- Electronic repair techs
- Warehouse personnel
- Mechanical repair techs
- Document controllers
- Laboratory technicians
- Assemblers
- Process planners
- Complaint coordinators
- Reliability engineers
- Salespeople

These people, and probably many others in your organization, need to have their training needs identified, planned for, and provided in order to satisfy the standard. And this activity must be supported with objective evidence—quality records. Many companies have personnel with unusual training needs. Such training is necessary to substantiate personnel qualification or certification. Sometimes we think of these people as performing special processes that require critical skills, evaluated and validated by an independent approval agent. These special processes are such that the quality or integrity of their results may not be effectively evaluated afterward, thereby necessitating the application of inherent controls while the process is actually being performed. Part of those controls will have to be exercised over the operator administering the process. One popular and practical control method is to rely on special, customized employee training, with certification following demonstrated competency through performance and/or testing. A few examples of these people/positions are the following:

- Welders/brazers
- Nondestructive test operators

- Plating line operators

- Heat treat operators

- X-ray technicians

Hopefully, by now you are realizing that virtually every person, every job, and every position in your company is affected by, or can affect, the quality discipline—and therefore is subject to paragraph 4.18 in the standard, Training. There are virtually no exceptions under QS-9000. Depending on the organization and scope of the quality system, ISO 9000 may permit a few exceptions such as for

- Field sales personnel

- Selected staff

- Union officials

- Financial personnel

- Facilities maintenance

You can readily see that the possible exceptions to the training requirement are few, if any at all. It may be, though, that not every aspect of every person's activity is governed by the standard. The task or challenge is to identify which aspects and which people.

All too often, companies take a limited/restrictive approach to setting the scope of their training program or initiatives. This is especially the case in traditional manufacturing organizations. Their first attention turns to machine operators and other production personnel. Secondly, they may address inspectors. Sometimes they'll get as far as some support functions like packers, field technicians, laboratory technicians, warehouse servers, calibration technicians, and equipment maintenance personnel. But as you can see, even these are traditional, operations-type functions. They're often referred to as hourly, associate, nonexempt, or blue badge employees. They are the heart of the operation side of the business. While satisfying the standard's training requirements for these types of positions is an excellent start, it does not go far enough. In all likelihood, there is another whole dimension to your workforce. You might refer to them as salary, exempt, nonunion, staff, or white badge employees. They are just as integral to the organization as the operations and line personnel in previous examples. And they are just as likely to affect, or be impacted by, quality. To get started, consider these few examples.

- Customer service reps
- Marketers and salespeople
- Department supervisors
- Internal auditors
- Support clerks
- Chemists
- Data analysts
- Computer programmers
- Trainers
- Order processors
- Laboratory assistants
- Artists
- Buyers
- Planning personnel
- Compliance officers
- Department managers
- Administrators

A few years ago, a television commercial used to declare something like "All aspirin is alike." NOT! Nor are all training programs and initiatives alike, nor should they be. Your company may have adopted one single training effort for all employees. This is not likely, though. A better, more common approach is to identify, plan, execute, and document training along some customized/tailored lines. It could be different by job type, personnel classification, department, salary versus hourly, exempt versus nonexempt, or union versus nonunion. Your particular approach may be to differentiate by some combination of these factors. Don't paint yourself into a corner. Stay flexible and creative. The way in which you satisfy requirements of the standard, and those of your company's quality system, does not have to be the same for all employees.

The frequency at which the program identifies, plans, conducts, and documents routine training may also vary widely according to the following factors.

- Management review cycles
- Individual people
- Subject matter
- Planning cycles
- Group versus individual training
- Performance review cycles
- Budgeting cycles
- Position

Special purpose, impromptu, or ad hoc training initiatives may occur at any time for a variety of reasons. Your training program needs to accommodate them. However, flexibility and practicality must prevail. Such training needs may arise from the following:

- Individual personnel
- New hiring
- Product modifications
- New equipment
- Procedure changes
- Performance deficiencies
- Corrective actions
- Management review results
- Company downsizing
- Employee requests
- Supervisor recommendation
- Promotions
- Nonconformance analysis
- After extended leave

- Company growth
- Transfers
- Process changes
- Quality system changes
- Probationary employees
- Customer requirements
- Preventive actions
- Reorganizations
- New functions/roles
- Audit findings
- Classification upgrades
- Skill upgrades
- Employee suggestions
- Complaint resolution

Companies tend to do a better job of identifying, planning, executing, and recording the more formal, structured training such as those determined during annual performance appraisals. They also do better when there's a special program for large groups, such as hazardous materials training. Unfortunately, the ad hoc or immediate needs are not managed as well. Sometimes they're not documented at all.

Then there are the deferred training needs—something that is identified as a need today, but that may not be met until sometime in the future. Here's an example.

Suppose the process engineers order a new piece of equipment. When it arrives, they find that it has an advanced logic controller for the computer numerical control programs. Also, the operator programming language is at a higher release level than previous equipment they purchased. It's learned that the equipment manufacturer offers a special training course, but it won't be offered again for four more months. Rather than expect the technicians to pick up the knowledge on their own, which they can do eventually, the department supervisor decides the formal course is the better approach. So today he decides they should take the course in the near future.

Here's the snag. The quality system has made no provision for effectively recording this type of need, nor for ensuring that it is ever satisfied. So guess what happens. Two months later, a new supervisor is assigned to the department. One of her first tasks is to prepare a new annual department budget. She is unaware of the training course, so naturally she does not budget for it. Course time rolls around, and the technicians don't go. Not only is there no money, the new supervisor has unwittingly scheduled the technicians for a special project during the same time when the course is offered. Too bad. Guess the techs will just have to figure it out on their own. Sure hope this doesn't negatively affect quality too much.

I've seen a few clever and very simple solutions for this oversight. One such approach is a matrix for each employee that identifies all of the core skills, experience, and educational requirements for his or her immediate assignment. As each requirement is satisfied, it is checked off, dated, and initialed or signed by an authorized agent. All additional training is also added to the matrix. Deferred needs are added at the time they are identified. They remain with an open status on the matrix until, sometime in the future, they are provided for. This matrix can be a great tool for planning, budgeting, and scheduling training. As completed, it serves as a quality record. The registrar will love it.

When I participate on assessment teams to evaluate the compliance and effectiveness of training, the subject of grandfathering almost always comes up. This is almost guaranteed for long-established companies and for those with a mature workforce. There needs to be a scheme for taking advantage of training (especially when it was not formally documented) that was provided prior to the adoption of the company's quality system. Employees may have been hired with undocumented skills, or they might have acquired necessary ability over a few years of performing their jobs. And then there is on-the-job training. We've all heard the all-too-familiar, "They've been doing that job for 20 years." Just serving time on the job does not equate to adequate training and competency.

Grandfathering is a line in the sand, a baseline, a starting point—after which the approach prescribed in the quality system will be implemented, enforced, and documented. It is a legitimate method to bring all existing personnel into the system where existing documentation may not substantiate their earlier skills, training, and acquired experience.

Imagine the following scenario. I'm the lead assessor and I've just asked to see your training records for an assortment of manufacturing operators. The human resources manager responds with what I've heard a dozen times.

"Well, we grandfathered everyone in about six months ago, so we really haven't had to do much training. But we can show you records for a HAZMAT course we conducted a month ago, and for an internal auditor refresher we had just last week."

"Great!" I say. "So then all of your employees are fully capable of performing their present assignments and, as such, they've been evaluated and approved by some authorized person, perhaps their supervisor?"

"Exactly!"

"Do you have any objective evidence that this official grandfathering act did indeed occur and that the supervisor officially attested to favorable results of the evaluation?"

"Not exactly."

"Do you suppose that the supervisor had a mental list of all the important skills that the employee should have been able to demonstrate in order to show competency?"

"Yes, certainly! She may even have had some notes written down."

"Do you mean that your grandfathering effort was really more than simply waving a magic wand over the workforce and instantly declaring them fully trained and qualified to do their respective jobs?"

"Of course!"

"Do you suppose that all supervisors could formalize this effort?"

"Yeah. Each designated supervisor or senior operator could probably prepare a checklist for each position, review each affected employee against the appropriate checklist, then sign and date the results. Any noted deficiencies or needed improvements could be the driver for remedial or future training. This completed checklist would become the quality record to substantiate the grandfathering. And we might even adopt the blank checklists as quality forms that can be updated as needed. What do you think?"

"What a great idea you have there! That would really make the grandfathering process meaningful. And from that date forward, you would follow your formal training program as described in the procedure?"

"Right!"

To recap, the standard wants your quality system to formally address training. It needs to be administered according to procedures, following policies in the quality manual. While it is a plus to accomodate informal, ad hoc, and spur-of-the-moment training, the system needs to approach training in a systematic (not bureaucratic) way. Required training must be identified, planned, conducted, and documented. Remember, the scope of your training program and initiatives must cover all personnel peforming activities affecting the Big Q.

4

ISO 9000 and QS-9000 Quality Systems: Platforms for Change

Change affecting your business is inevitable! There will be changes in customers, customers' needs and expectations, market conditions, employee skills, manufacturing processes, products, capital and resources, competition, subcontractors and vendors, quality requirements, organization size and structure, raw materials and their availability, equipment capability, costs and investments, and change in technology—just to name a few. Change can be either positive or negative, welcomed or not—but it is still inevitable. Change can be ignored, resisted, injurious, even fatal—or it can be managed for success.

Documented and effectively implemented quality management systems provide a platform upon which change can be constructively managed.

The ISO 9000 series requirements standards (and the automotive version, the QS-9000 requirements) are excellent quality systems standards that provide models for quality assurance in design, development, production, installation, and servicing. These standards include 20 elements, or planks, for building your company's own platform for change. Each plank describes guidelines and requirements for defining and managing nearly every aspect of your operation, offering the tools for constructive, successful change. Quality management systems accommodate change. There are upside changes everyone welcomes, such as

- New products and processes
- Resource additions
- New technologies
- Skills development
- New/expanding markets
- Failing competition

Then there are the dreaded downside changes, such as

- Employee turnover
- Resource reductions
- Diminished production
- Demanding customers
- Escalating costs
- Product obsolescence
- Statutory requirements
- Process changes
- Loss of critical skills
- Consolidations
- Increased competition
- Loss of key suppliers
- Regulatory codes
- Government controls

Let's take at a look a number of requirements from the ISO 9000 standards and QS-9000 requirements to see how compliance to them can result in the capacity to successfully manage change.

Defined and Documented Quality Policy and Quality Commitment

Adopting, documenting, and communicating the company's quality policy and commitment statement(s) is an excellent first step. It must be followed with evaluation to ensure that all personnel, throughout all levels of the organization, really do understand them. Regular reinforcement, combined with genuine and sincere demonstration by management, is vital. This will provide a stable and consistent theme and focus that can guide quality and related activities through difficult business times. Think of the quality policy as the compass, and the commitment statements as the rudder and keel. They don't seem all that important when the sloop is safely in the slip or moored at the marina restaurant. But when having to tack upstream, or overcome a squall eight miles out, you may find there is little else you can depend on.

Defined and Documented Quality Objectives

Clear definition of measurable quality objectives (with associated qualitative and quantitative targets) will help ensure meaningful effectiveness and suitability assessment of the system. They will help management maintain its objectivity. And such measurables can enable your quality system to evaluate the positive results or negative consequences of change—not unlike the ship's barometer. They can serve as the engine gauges (oil pressure, water temperature, tachometer) of the quality system by monitoring vital performance characteristics.

Formal Management Reviews of the Quality Management System

Formal, documented reviews of the system by executive-level management will ensure the ongoing suitability and effectiveness of the quality program. This forum provides an excellent audience for assessing progress against the quality objectives. It is important that the reviews occur at a prescribed frequency, follow a core agenda, and have a selected mandatory attendance. If so, these reviews can ensure ongoing evaluation for improvement, despite pressure to be distracted by daily issues (fire fighting), while commanding appropriate management attention. Can you imagine the skipper leaving the helm just as the ship is coming through the channel, distracted by a squabble among the crew?

Documented Responsibilities, Authorities, and Interrelations

For those people within the organization that perform functions affecting Big Q (not just the quality department or product quality control), there shall be clearly defined and documented responsibilities, interrelations, and commensurate authority. This will ensure that important activities are appropriately assigned and that adequate resources are provided. Key duties are then much less likely to fall through the cracks if and when resource consolidation occurs. Nonessential or redundant efforts may be exposed for possible reassignment.

Documented Quality System Procedures and Work Instructions

With the quality assurance manual as the top-level umbrella document, the quality system must be thoroughly described in a network of subordinate procedures and work instructions. Their content and level of specificity may vary depending on the complexity and criticality of the work, the experience of personnel, and actual training. This documentation will define the proper and preferred way of doing things, from administrative tasks to engineering initiatives to machine operations. When personnel change through attrition, transfer, or promotion, activities can continue to be performed consistently.

Consistency aids in the optimization and streamlining of administrative or operational processes. By having to think through a process before committing it to documentation, process owners have the incentive to identify and plan for essential components. There may be opportunities to streamline the process by exposing redundant or unnecessary activities.

Documented Procedures for Quality Planning

Disciplined, procedural quality planning can, in a cross-functional environment, provide the methodology for establishing and maintaining consistent product quality. It is accomplished by identifying and satisfying all internal and customer requirements, by minimizing variation in quality control activities should personnel change, and by identifying and controlling important product and process characteristics. Such quality planning is a proactive, preventive exercise. It causes the organization to do upfront thinking, rather than reacting afterward when those unplanned failures occur.

Many companies are frustrated by having to redo so many of their core activities, or by the excessive resources consumed by the most structured tasks. For them, daily production or quality meetings have become little more than obsessions with compiling the priority list of fires to fight and crises to overcome. Time was not taken for genuine planning beforehand, so they must continue to respond to conditions that have not been placed under control—either before the problem was manifested, or at least while it still had manageable proportions. Formal quality planning will help to stop the insanity. At first, incremental resources and time may be necessary for this effort, but eventually reductions can occur. Savings in time, people, and equipment can be diverted to productivity. In simple terms, you either plan or you try to recover. You either prepare or you suffer the consequences. Quality planning is a choice.

Documented Reviews of Contract Requirements

Temper the compulsion of the marketeers to take on any and all business. Has your operation ever been the unsuspecting victim of misunderstood or unknown customer requirements? Or, on occasion, has your customer been the victim when you failed to follow through with your obligation? Perhaps your failure was not intentional. Perhaps the customer's requirements were not clear to you. Formal, documented reviews of contracts will identify all customer requirements, eliminate ambiguities, and aid in resolving exceptions. The reviews should encompass all contractual components—purchase orders, drawings, specifications, supplements, and amendments.

Properly done, contract reviews will reduce your occurrences of scrap and rework; returned goods; customer complaints; delayed shipments; wasted time, materials, and labor; and unprofitable orders, to mention a few. Knowing the requirements and being assured of your ability to satisfy them will help keep costs down and customer satisfaction high. This is essential in times of undesirable change.

Formal Preventive Action Activity

Regular, systematic analysis of quality information for proactive and preventive opportunities will develop and improve the operation. It can become more robust, productive, effective, and efficient by reducing cost and eliminating wasted personnel and materials. Investigating preventive opportunities can capture individual and team creativity and best practices. Results can avoid some problems altogether and lessen the impact of others. Routine prevention activities keep the patient well, making the company more resistant to illness associated with negative change.

Root Cause Analysis and Corrective Action Implementation

Thorough root cause analysis and effective implementation of resultant corrective actions can eliminate future recurrence of failure. Proper, detailed analysis can identify the culprit so that precious resources will be properly targeted. The shotgun approach is far too wasteful and unpredictable. Missing the mark by shooting down the wrong alley will consume unrecoverable time. Taking purposeful, deliberate corrective action that is later verified to ensure actual implementation and effectiveness will go a long way toward resolving the problem so that it won't be repeated. How many of your current problems have been around for a while? How many times has your company fixed them, and fixed them, and fixed them again? Externally imposed change will demand the full attention of your quality system. It must not be diverted by repeating past mistakes. By having well-defined and documented procedures for identifying, implementing, and verifying corrective action initiatives, you'll have another plank in the platform for change. Extra personnel can be freed to work on meaningful, productive issues.

Identifying Needs and Providing Training for Personnel

For all personnel performing functions that affect Big Q, it is essential to identify their training needs, provide that training, and then document the results. Doing so can help the system capture and reproduce vital knowledge and skills. It can help ensure correctness and consistency in all activities. In times of shrinking resources, it is appropriate that personnel be capable of performing multiple functions, in more than one role. Proper training can facilitate this. More effective cross-utilization of people can result. Your operation can make the most of its skill base, despite employee loss or transfer.

Key Product and Process Control

Identifying and controlling key process parameters and product characteristics must be done if costs are to be contained and outgoing quality improved. Appropriately qualifying processes, equipment, and personnel can optimize processes to reduce variability, curtail wasted materials, and minimize rework and scrap. Productivity and efficiency can result. Ultimately, customer satisfaction occurs. Conditions rarely improve on their own. Process capability and product quality never improve without intervention. Processes must first be understood: what is important versus what is critical. What impacts what? What are the contributors? Once understood, the process can be done consistently and then brought under control. Only then can significant improvements be realized.

Identification of Procurement Requirements

Thorough identification, documentation, and communication of procurement requirements can ensure improved quality of goods and services from vendors and subcontractors. By eliminating incomplete, ambiguous, or contradictory requirements, your suppliers will be better informed and equipped to satisfy your needs. Improved vendor relationships can result. Reviewing purchasing data prior to release will cause responsible people to identify requirements that have been overlooked or not correctly described. Once addressed, getting what you really want becomes more likely.

Use of Evaluated, Approved, and Controlled Subcontractors

Effective selection, monitoring, and control of subcontractors and vendors can help provide quality goods and services, improve consistency, and promote prompt deliveries at reasonable prices. There may be opportunity to maintain less inventory and to have fewer receiving inspection activities. Costs associated with nonconforming product and delayed deliveries will be reduced. Emphasis on receiving inspection can be diverted to value-adding activities. Negative effects on your own internal processes can be controlled by eliminating nonconformances and reducing variation in purchased goods.

Let's conclude by briefly restating some of the structural planks in your platform for change.

- Defined and documented quality policy and quality commitment
- Defined and documented quality objectives
- Formal management reviews of the quality management system
- Documented responsibilities, authorities, and interrelations
- Documented quality system procedures and work instructions
- Documented procedures for quality planning
- Documented reviews of contract requirements
- Proactive preventive action activity
- Root cause analysis and corrective action implementation
- Identifying needs and providing training for personnel
- Key product and process control
- Identification of procurement requirements
- Use of evaluated, approved, and controlled subcontractors

5

The Menu of Quality Objectives and Policy

This chapter features fictitious sample quality policy statements, organizational goals, and many examples of associated, realistic quality objectives. Throughout is an overwhelming list of practical examples to drive home the truth that you really can evaluate the health and wealth of your company's quality system regardless of its degree of sophistication, formality, and level of maturity. They are in no way intended to be prescriptive or mandatory—just something to get your creative juices flowing. No doubt, you'll be able to add many more examples from your own company. The challenge is to select those most appropriate for your particular quality system. This chapter is an excellent place to begin.

Let's consider a hypothetical sample quality policy and associated organizational goals and objectives.

ACME Basket Company is committed to excellence in all we do and to providing the highest-quality goods and services to our customers. We will strive to

- Offer new and innovative products to the marketplace.

- Consistently meet or exceed customer specifications.

- Deliver quality products on time and at competitive prices.

- Ensure outstanding customer satisfaction.

- Be number one in our industry.

To do so, we will

- Purchase only the best raw materials.

- Continually improve our process capability (C_{pk}).

- Develop the skills of all employees.

- Involve personnel in problem solving.

- Encourage employees to contribute in system improvement.

- Utilize proven quality tools.

- Monitor and respond to customer perception.

- Regularly benchmark and conduct competitive analyses.

- Optimize administrative processes.

You can see that the "to do so" list lends itself very nicely to measurable, qualitative, or quantitative attributes.

The standard specifies objectives for quality in paragraph 4.1.1 (quality policy) and refers to them again in paragraph 4.1 (management review). It is the quality system's success in satisfying these objectives (among other things) that indicates the system's suitability and effectiveness. It is these objectives for quality that must be evaluated in the management review. More correctly, it is the quality system's performance against them that must be evaluated. We should have a clear understanding of what quality objectives are and of some of their characteristics. Listed here are a number of the attributes of quality objectives. This should stimulate your creative thought process.

Quality objective attributes

- Reflect on the system's suitability and effectiveness.

- Relate to components of the quality policy.

- Address the organizational goals.

- Should be qualitative and/or quantitative.

- Must be defined and officially documented.

- Are constant categories with escalating targets.
- Include special purpose and limited-time categories.
- Are attributable to the quality system.
- Include realistic/stretch goals.
- Relate to customer wants, needs, and expectations.
- Address negative customer perceptions.
- Are supported by lower-level goals, objectives, and measurables.
- Link to quality system activities and processes.
- Facilitate evaluation and measurement.
- Are unambiguous.

To fully appreciate quality objectives, let me share some examples. Keep in mind that these particular ones are not mandatory or prescriptive. They may not even be appropriate for your system. Again, you are challenged to find the right ones for your operation and quality system. The better the selection, the more they'll tell you about the performance of Big Q in your company.

Here are some examples of quality objectives.

- Improve the customer satisfaction index by three points.
- Achieve a process capability index of 1.33 for key characteristics.
- Reduce scrap and rework costs by 10 percent annually.
- Prevent further escalation of the late delivery curve.
- Reduce appraisal costs by $2K per month.
- Increase prevention investments to $1K per week.
- Improve product development launch time by 10 percent per quarter.
- Include production operators on corrective action teams.
- Develop a dock-to-stock supplier program.*
- Transfer process inspection tasks to certified operators.
- Reduce internal engineering changes by 20 per month.

- Reduce time for order taking and entry processing by 15 percent.

- Eliminate duplicate suppliers on nonessential materials.

- Reduce nonmanufacturing lead time to 10 percent of throughput time.

- Maintain WIP inventory costs below $2.3 million.

- Obtain top quality awards from three key customers.*

- Improve product audit pass rate to 98 percent.

- Reach cross-training utilization of 40 percent.

- Introduce three core product models per year.

- Realize manufacturing throughput time of 32 shifts maximum.

- Improve machine available up time to more than 90 percent.

- Achieve ISO 9000 certification at all six sites in 36 months.

- Become a sole-source supplier to our top five customers.*

- Increase market share to 30 percent.

- Grow market share at a rate of 3 percent per year.

- Achieve training goal of 32 hours per employee per year.

A couple of brainstorming sessions including the quality system experts in your company should yield a list of 20 to 30 possible quality objectives. But do not fall into the trap of measuring the system against too many. You will be better off selecting the very best, most revealing, and representative quality objectives. In the better systems, I typically see five to 10 that are clearly defined, officially documented, and measurable. With the previous examples as a springboard, you should never again be at a loss for attributes or characteristics by which you can qualitatively or quantitatively evaluate the effectiveness of your quality system. You should never again be at a loss for quality objectives.

Some quality objectives may actually be organizational goals; see the items marked with an asterisk (*) in the preceeding list. Some may be expressed in qualitative terms, while others may be quantitative. They are often expressed as attainment of operational parameters, reduction in failure rates, or improvement in negative trends. In any event, they must be clearly defined and documented. You must be able to compare system performance against them. They must establish real expectations for the system. They must tell you whether quality matters are declining, remaining status quo, or improving. It's been said, "If you don't know where you're going, any road will get you there." Put another way, "You may be lost and just don't know it."

The quality system in your company needs a survival kit. It needs a compass, map, flashlight, string, knife, snakebite kit, food rations, and matches. It must be equipped to spend a couple of nights in the dark woods if need be. It needs a quality policy and objectives that are

- Clearly defined

- Realistic

- Attainable

- Growth oriented

- Reviewed by management

- Documented

- Qualitative

- Monitored

- Communicated

- Measurable

- Quantitative

- Updated

- Reflective

Let's use the analogy of your annual health checkup and physical. Would you be satisfied if the physician merely visually examined you and declared, "You must be in good health [suitable and effective quality system] because you have a great tan, you don't appear to be overweight, you have a wonderful smile, your posture is good, and gray hair hasn't yet set in"? Of course not! Although all of us would like to hear this feedback from our doctor, it may only give a hint as to your true physical well-being. To draw conclusions based solely on this visual critique would be highly subjective to say the least, if not outright incorrect. Judging your health by this type of cursory exam is no more effective than to declare your quality system to be suitable and effective based on observations that sales are increasing, employee morale is high, profits are up, and the number of customer complaints is decreasing. These superficial indications could just as easily be the result of failing competition, the new employee cafeteria and health facility, industry-driven price increases, and customers who are tired of complaining and not getting any satisfaction.

To effectively assess the condition of your body (quality system), the physician has learned that a dozen or so key physical attributes (quality objectives) must be evaluated. These will include such things as blood pressure, heart rate, cholesterol levels, body fat content, reflexes, blood and urine analysis data, hearing and sight test results, and so on. These objective, clearly defined attributes will have established targets or ranges that indicate whether the patient is healthy or not. Results outside of the preferred range may indicate possible health problems. At the least, a closer look is warranted.

Your company's management review team needs to "play doctor" with your quality system. Specific health attributes and vital signs need to be established that can serve as criteria against which the system's suitability and effectiveness can be evaluated during the management review. In the absence of these quality objectives, the team may be tempted to declare that the system is in great shape because of the good weather, the new company lease cars, remodeled offices, and other things that are nice to have, but that do not truly reflect on system performance.

For the "motor heads" out there, here's another brief example. Would you say your engine is in tip-top condition just because it *sounds* good? Certainly not! (Maybe yes, if you are not a genuine motor head). Rather, you have set performance expectations for the motor, such as compression, manifold vacuum, oil pressure, water temperature, torque to rpm ratio, and so on. When actual values fall within range, all is well. If any values are out of limits, it's a good bet that that the engine is in need of adjustment, repair, of maybe even an overhaul.

6

Considerations for an Effective Management Review

For the past few years, we've heard a great deal about the global marketplace, world-class quality, corporate downsizing, reengineering, TQM, and economic control by the European Community. For most companies, the solution of choice to these challenges has been to adopt a formal quality management system. Very soon, these companies came to realize, regrettably, that simply having a system does not ensure its effectiveness or positive results. I've heard some disparaging words about TQM for this very reason. The naysayers have some legitimate gripes. But I don't believe in attacking the system. The principles are sound. The blame rests with the implementation and nurturing of the system, and with managers for not understanding and taking ownership for the quality system. They haven't taken appropriate steps to ensure the system's success.

Management review of the quality system is perhaps the single most powerful tool for leading your company through change and toward improvement. This executive review will help to ensure continuing suitability and effectiveness of the quality management system. It causes management to address probing questions about the system. By establishing and monitoring meaningful quality objectives, the management review can further measure system performance.

Management review will be approached as a process, with discussion on its inputs, actions, and outputs. At the end of the book, the minutes from a real-life management review (compliments of Saft America, Industrial Battery Division) will be offered as an excellent example for your consideration. It is one of the best I've had the pleasure of assessing.

Imagine that you are Captain James T. Kirk, Commander (plant manager, chief executive officer, and so on) at the helm of the U.S.S. Starship Entrepreneur. Your mission: Within six months, take your ship and crew (your company and its quality system) where no one has gone before—to a far distant galaxy known as EMR (excellent management review). Have your navigator lock in the course coordinates. Alert your chief engineer (management representative) to bring the engines up to full power. Lower your defense shields (paradigms and turf barriers) and proceed ahead at warp 9! "But Captain, will the ship take it?" "It must, Scotty—our survival (quality and profits) depends on it. Make it so!" "Aye, aye, Captain."

The ISO 9000 series requirements standards contain specific requirements about management reviews. They state in paragraph 4.1.3,

> The supplier's management with executive responsibility shall review the quality system at defined intervals sufficient to ensure its continuing suitability and effectiveness in satisfying the requirements of this International Standard and the supplier's stated quality policy and objectives (see 4.1.1). Records of such reviews shall be maintained.

In a closely related paragraph, 4.1.4, Quality Policy, it says,

> The supplier's management with executive responsibility shall define and document its policy for quality, including objectives for quality. The quality policy shall be relevant to the supplier's organizational goals and the expectations and needs of its customers.

In paragraph 4.16, Control of Quality Records, it says that "Quality records shall be maintained to demonstrate conformance to specified requirements and the effective operation of the quality system."

At the beginning of this chapter I referred to probing questions about the quality system that should be addressed by management.

The purpose of the management review is to answer a number of these difficult questions, including: Why do we have it? What do we expect it to accomplish? Is it still suitable for our needs? Can we evaluate its effectiveness? Can it, or should it, be changed/improved? Are we realizing benefit from it? Is it meeting the ISO 9000 standard? Does it help us satisfy our quality policy? Is it helping us reach our organizational goals? Does it allow us to satisfy customer needs/wants? If the management review team is unable to answer these questions, the quality system is not well understood, and management does not understand how it works or what it is expected to accomplish for the company. Essentially, the system is not being managed!

The standard requires the management review to evaluate the continuing suitability and effectiveness of the quality system. To appreciate this requirement, let's begin with a discussion on suitability, then move on to effectivity. You will find it somewhat difficult to distinguish between suitability and effectiveness for a few of my examples and for some of your own. That's the reality of it.

Consider system suitability. Think about all of the various activities prescribed by the standard, and by your own system documentation. Now focus on the top ones (maybe 10 to 20, depending on your particular company). Focus on those that consume the most time and resources—those that are the most critical to your processes and products. Be bold enough to compile a list and take it to your next quality system management review. Then be really radical! Challenge the team members to decide if various aspects of those activities are still suitable today. Can they defend their decisions?

To get you going, I've made up my own list. Use it as a starter. Add or delete items as necessary. It doesn't have to be perfect or all-inclusive the first time. I'm positive it will be changed the first time it is critiqued by the review team. And it will be refined as the management review process matures and as the quality system evolves.

Are various aspects of the system still suitable? What aspects of suitability are we talking about? Consider critiquing virtually every facet of the system. Approaching it element by element according to the standard is the most common method. Dealing with it department by department, product by product, or even process by process are other approaches. Consider the following examples. Challenge the way an activity is currently being performed against possible alternate methods.

- Receiving inspection versus vendor control
- Final inspection versus process control
- Inspectors versus certified operators
- Frequency and focus of internal audits
- Permanent versus subcontract employees
- Organizational structure
- Resource allocation and distribution
- Level and intensity of operator training
- Detail and content of work instructions
- 100 percent inspection versus statistical sampling
- Internal versus external training
- Centralized versus localized document control
- Vendor audits versus ISO 9000 registration
- Frequency, focus, and agenda of management reviews
- Divisional versus plant site activities
- Make versus buy and outsourced services
- Quality planning by project versus process versus product
- Handling, storage, and preservation of in-process materials
- Hard copy documentation versus online, paperless system
- Method of conducting contract and specification reviews
- Method of carrying out quality planning
- Process for performing design reviews and validations
- Process for documenting and resolving customer complaints

Consider system effectiveness. Repeat this exercise on suitability, except this time think about efficacy of the quality system. Make your list, take it to the review, critique the activities and aspects, then revise the list for next time. Following is a starter selection. Management review should ask: Are various aspects of the quality system operating effectively? The team should look at the way the company performs the following activities.

- Identifies and traces materials and products
- Controls and disposes of nonconforming products
- Collects and files quality records
- Verifies corrective and preventive action implementation
- Identifies and investigates preventive action opportunities
- Solicits and evaluates customer feedback
- Determines and controls key process parameters
- Identifies and measures key product characteristics
- Identifies, applies, and controls statistical techniques
- Performs and documents results of contractual service
- Selects, approves, and controls vendors
- Sets measurements of effectiveness
- Establishes goals and objectives
- Revises, approves, and distributes controlled documents
- Performs multifunction contract reviews
- Develops and monitors design project plans
- Specifies and reviews vendor purchasing requirements
- Selects and applies sampling plans
- Utilizes roving floor inspection personnel

Earlier, I raised a few probing questions that you may wish to pose to your management review team. But first, let me arm you with a few of the answers. These are typical responses, but may vary considerably from the answers given by your management team.

Why do we have a quality system?

- To have consistent methods and practices
- To communicate internal and external requirements
- To ensure high-quality goods and services
- To reduce and simplify customer audits

- To improve operating performance and profits
- To distinguish ourselves from our competition

What do we expect the quality system to accomplish?

- Reduce nonquality costs (scrap, rework, repair, and so on).
- Provide correct and timely documentation.
- Minimize customer complaints and returned goods.
- Reduce nonconforming purchased goods.
- Eliminate unacceptable vendor services.
- Control or reduce inspection/appraisal costs.
- Increase productivity and efficiency.
- Improve worker utilization and cross-training.
- Limit product development and launch times.

What measurements can we use to evaluate system effectiveness?

- Cost of off-quality goods and services
- Customer satisfaction data
- Vendor performance results
- Internal and customer audit findings
- Operator efficiency ratings
- Nonmanufacturing lead times
- Equipment downtime and repair costs
- First-time-through ratios
- Number of cross-trained operators
- Time to process engineering changes
- Frequency of customer deviation requests
- Product audit pass ratios
- Process capability indices
- Number of preventive actions implemented

- Time from design concept through validation
- Volume of returned goods authorizations
- Inventory cost of WIP
- Production cycle times
- Prevention-to-detection cost ratio
- Percent of regraded product
- Closure time for corrective actions
- Recurrence rates for defect codes
- Frequency of documentation changes
- Final inspection acceptance rates

You should never again be at a loss for attributes or characteristics by which you can evaluate (qualitatively or quantitatively) the effectiveness of your quality system.

The Management Review Process

By my definition, a *process* is an event or exercise that takes input; acts upon it, causes a conversion, or otherwise adds value to it; and then produces an output. As promised earlier, I want to approach the management review meeting as a process. Accepting my definition of a process, what are some of the important process components for management review? As a minimum, you should ask yourself: Who should attend the meeting? How often should the review be held? What are some of the key responsibilities that need to be assigned? What are the essential input data for the review? What actions and considerations will take place in the meeting? What valuable outputs should the review meeting produce?

There are no single right or wrong answers to these questions, but they must be asked, and your answers must be correct and reflective of your organization. Let's examine each component, and I'll offer a dozen or more potentially acceptable responses to these questions. From my lists, you should be able to pick a few answers that are right for your company. Your knowledge and experience will permit you to expand your own list.

Keep in mind, management reviews are highly individualistic to a particular company's quality system. Just before I learned that little girls were "sugar and spice and everything nice" (that would grow up to be God's best creation), I was captivated by the hobby of building plastic model cars. Occasionally I'd build it "stock," but my heart was into customizing. One particular model kit manufacturer used to provide extra grills, front and rear end profiles, tail lights, and a wonderful little tube of plastic body filler. To these extra components and materials I would add small swatches of fabric, glitter, metallic paint, and lots of time. The results were something magical to behold. My cars had body styles like no other kid's cars. Every one was unique and my customized creation. The wild colors and velvet interiors made my beauties winners in almost every contest. Saturday mornings found the boys lined up in front of the five and dime store (though even in the mid-1960s you couldn't buy much for a dime, let alone a nickel). We'd rush in as soon as the owner unlocked the door and, like magnets, we were drawn to the model car display case to see whose entry won first prize. Looking back, and knowing that the store owner and his wife chose the winners, I'm sure that nearly every guy got a chance to win a ribbon eventually. My son never experienced the thrill of model car building; his heart was stolen by more modern things like radio-controlled model cars and video games. As it is with kids and model cars, so it is with management reviews—individuality matters.

Who should attend the management review? It is crucial that the proper people attend and participate. Remember, this is not just the quality department's program—it is a quality management system for the entire organization. And the whole organization stands to reap the rewards. We need to have decision makers there. The standard requires that managers with executive responsibility attend. Those with a vested interest in the quality system should attend. The ISO management representative is vital. He or she is responsible for ensuring that the system is established, implemented, and maintained, and for reporting on its performance. Some attendees may only be invited observers. Following is a menu from which you can select your attendance roster.

- Middle and senior management and staff
- Quality steering committee
- Material review board
- ISO management representative

- Special designees from the quality department
- Corrective action assignees
- Key union representatives
- Special customers
- Special subcontractors, vendors, and suppliers
- Functional department heads
- Key project leaders
- Preventive action initiators and champions
- Quality manager

How often should we meet? The frequency of the review meeting is important. The standard requires that it be performed at defined intervals. The frequency may change over time, depending on the system's particular structure and needs. It should not be so frequent that it gets bogged down in the minutia of everyday operations. Nor should it be so seldom that it loses management's attention and the system goes wandering. The frequency should be appropriate to permit good leadership. This is a high-level, "let's stand back and take a look at the big picture" kind of event.

When I was growing up, I loved to be around my dad. Dad had several creeds by which he lived. One was, "Never pay someone else to do what you can do yourself." I now realize that raising five children on a middle-income salary probably had some influence on Pop's approach to automobile repair. He taught me almost everything, from changing the oil to regrinding the valves to dropping a "tranny." At the start, I regularly asked, "How tight should I tighten this bolt, Pop?" (we didn't own a torque wrench then), to which he always replied with a smile, "Just enough, my son." After a couple of broken studs and many busted knuckles, I learned what he meant. So it is with the frequency of management reviews. It should be just enough—not too seldom, not too often. They don't make a torque wrench for this task, so let me suggest a number of other indicators that will let you know what is just enough. Ponder on these knuckles, if you will.

- It may depend on the maturity of the quality system.
- It may depend on the results of the previous reviews.

- You may require interim, special purpose sessions.

- Consider changes to the system and company.

- It may be timed to follow lower-level reviews.

- They are typically quarterly, semi-annually, or yearly meetings.

- Not too seldom, not too often—meet just enough.

Quality system processes never happen on their own. They have to be implemented through people. People make quality happen. On assessments I sometimes wear a campaign-type badge that declares, "Quality Happens Through *People!*" Seeing the badge helps to elicit pride and confidence in the auditees. People make the management review go. Though not a requirement, I think it is a good idea to have the ISO management rep handle many of the logistics. The standard does say that he or she must report to management on the performance of the quality system. The representative needs to assign key responsibilities for the event.

What are the responsibilities associated with planning and executing a management review? The answer will depend on your company's size, organization, culture, formality, and personnel disciplines. But, as a minimum, the following should be addressed.

- Set and publish an agenda with core and special items.

- Secure adequate resources and facilities.

- Identify and notify special invitees.

- Notify regular attendees of time and location.

- Collect and distribute pertinent input data.

- Follow up on action item assignments.

- Chair and/or facilitate the meetings.

- Document and publish the records and minutes.

What are the inputs to the management review? To qualify as a process, according to my definition, the management review must receive input. The kind of input can vary greatly and, therefore, needs to be determined ahead of time by the people who need to review it. The format and the level of specificity of the data is important. Don't forget who will be attending or why they are meeting. Stacks of computer printouts, a two-hour slide show, lengthy speeches, and dozens of graphs may not be what the attendees had in mind. Summary data and reports and a few key overheads with succinct oral explanation will communicate the big picture just fine. But do have the supporting detailed information available as back-up. With the following list as a menu, you should never again be at a loss for what to bring to the management review meeting. Don't be overwhelmed; it's only a wish list. As you read it, keep this in mind.

Pretend you're a kid in a candy store. You only have 87 cents and three empty pockets, so you can't buy every piece of candy on the shelf. Pick the best ones. Choose those that bring the most satisfaction. Maybe experiment with one of those weird orange and purple balls; they may prove to be everyone's favorite. Don't reject a certain kind just because you haven't tried it before. Sweet isn't always best either, try the sour ones too.

- Summary data from lower-level reviews
- Cost of quality data
- Training needs and activities
- Competitive analysis information
- Benchmarking results
- Performance metrics
- Customer survey responses
- Core agenda items
- Information on key productivity issues
- Special project updates
- Customer presentations
- Quality audit results
- Overdue corrective action responses
- Process audit results
- Preventive action information
- Process yield statistics
- Productivity ratios
- Scrap, rework, and repair statistics

- System element issues
- Vendor presentation material
- Market data
- Personnel resource issues
- Key supplier data
- Vendor performance reports
- Employee suggestions
- Customer complaint and return data
- Customer audit results
- Third-party audit results

What actions should be considered? A process must act on, convert, or otherwise add value to the input if it is to produce worthwhile output. For management review, the conversion comes in the form of discussion, thought, consideration, evaluation, and determination (on or about the input). What are some of the issues that can be considered in this conversion? What should the review team members be asking about the input data?

- Does each metric meet its target?
- Is the quality policy still relevant?
- Are quality objectives real and attainable?
- What are strengths and weaknesses of each system component?
- Are we progressing toward each goal at an acceptable rate?
- Is there evidence of true preventive/proactive activity?
- Are pertinent projects on course?
- Should targets be increased or decreased?
- What changes are needed, why, when, and how?
- Are new organizational goals appropriate?
- Is our commitment to quality actually manifested?
- Are customers' needs and expectations being satisfied?

What are the outputs from the management review? Every process must have a reason for being. There must be an end for the means. For management review, its purpose and its output must be conclusions, decisions, and summaries about the suitability and effectiveness of the quality system. Just collecting and poring over a bunch of facts, figures, and graphs does not make a management review. It's what the information tells you, what is learned, what judgments are made, and what is done as a result that is the substance of the process. What tangible, documented evidence should there be in order to demonstrate that a truly meaningful, effective management review has taken place? What is the output that separates this meeting from the hundreds of other unproductive meetings that burden companies? What reward makes the effort so worthwhile? Can your management review team point at something and say, "We accomplished that"? The best management reviews produce a majority of the following outputs.

- Documented minutes and reports
- Action items and assignments
- Recommended system changes
- Corrective actions
- Identified strengths and exposed weaknesses
- New/revised goals, targets, and objectives
- New/revised resource allocations
- Revised internal audit programs
- Dates and responsibilities for follow-ups
- Preventive actions
- Improvements in the review process itself
- Changes to the review's input data
- Revised lists of review meeting attendees
- Judgments, conclusions, and determinations
- Final summaries of the reviews
- Lessons learned

Congratulations! You, your ship, and your entire crew have arrived safely on "Great Day," the twenty-second planet of the third sun in the EMR galaxy. The inhabitants are a highly intelligent species. They have been waiting 72 light-years for someone with your wisdom to lead their federation's quality management system.

7

Considerations for Effective Contract Reviews

The requirements for contract review, as discussed in this chapter, have been taken from the international standard ISO 9001, *Quality systems—Model for quality assurance in design, development, production, installation and servicing.* Comparable requirements are also found in ISO 9002 and QS-9000. Element 4.3 of the standard deals with contract review and includes four subclauses: general, review, amendments, and records.

Let's begin with discussing the general requirements. The standard says that the supplier (that is, your company) must establish and maintain documented procedures that describe the process for reviewing contracts and for the coordination of those review activities. There is also a reference note in the standard that strongly suggests that you establish (define, document, and implement) communication channels (internal and external) and interfaces with your customer's organization. Because this is mentioned in a note, it is not mandatory. Nonetheless, it should be done if you desire excellent contract reviews.

What is meant by *coordination of activities*? Consider the following:

- Who champions the contract review process? (for example, quality)

- Who orchestrates the contract review process? (for example, marketing)

- Who participates in the process?

  —Engineering

  —Production

  —Manufacturing

  —Quality

  —Marketing

  —Research and development (R&D)

- Who interacts with the customer?

  —Marketing

  —Field service

  —Quality

  —Applications engineering

  —On-site personnel

Coordination of activities suggests that someone or some group will be tasked and authorized to interact/interface with the customer and with those affected people/groups within the supplier's organization. Effective reviews will not consistently happen on their own; they must be administered and nurtured. The quality manual should acknowledge the requirement for management and provide a brief system-level description. Your documented procedures and/or work instructions must add the detail and address who, what, when, why, where, and how. They should describe various personnel/position/department authorities and responsibilities essential to the review process. Documentation should cross-reference other pertinent procedures and work instructions, and either identify or include necessary forms. Quality records of the contract review need to be specifically (not generically) identified.

In the very best (effective and thorough) contract review situations, there is cross-functional participation throughout the supplier's organization. It should not be just quality's job, nor surrendered entirely to marketing. Involvement by all affected and responsible departments, and possibly by those with simple interest, will ensure more effective, thorough reviews. In general, for any requirement, the person/group/department that is either most knowledgeable about, the most impacted by, or authorized to make decisions on the item should be involved.

Exactly what kind of review is appropriate will depend on several factors.

- Customer requirements
- Size and complexity of the supplier
- Type of products
- Supplier's organizational structure
- Industry expectations
- Nature of processes
- New contract versus amendment

The degree, extent, and complexity of the review; who participates; and the amount and type of documentation and quality records will vary. Obviously, a supplier making missile guidance systems will have a contract review process that is far more technical, complex, detailed and thoroughly documented than that of a company making pocket radios or men's dress shirts.

Consider the varying degrees of contract review and of those people or groups involved. It is unlikely that a supplier will always do the same level of review, by all the same people, and produce the same documentation and records for every case. The approach will probably depend on which of the following order types is being addressed.

- Simple catalog orders
- New or revised specifications
- Product modifications
- Repeat orders
- Process changes
- New/complex projects

The next clause of the standard (para. 4.3.2) deals with the review. There are two terms in the clause that need to be defined.

1. *Tender.* Offer made by a supplier in response to a request for quotation or an invitation to satisfy a contract award to provide the specified goods or service (proposal, quotation, and so on).

2. *Contract.* Agreed-upon requirements between a supplier and its customer, which may be transmitted by any means (verbal, written, electronic transmittal, telephone/fax, and so on).

Paragraph 4.3.2, Review, says that *before* a tender is submitted to the customer and *before* a contract or order (statement of requirements) is accepted by the supplier, the tender, contract, or order shall be reviewed to ensure that

- Requirements are adequately defined and documented.
- Verbal requirements are agreed upon before acceptance.
- Differences between the tender and the contract are resolved.
- The supplier has the capability to meet all requirements.
- Exceptions to acceptance are documented and resolved.

Why is the standard insistent upon review before submittal of the tender and acceptance of the contract and amendments? Simply because either the supplier reviews, plans, and functions intelligently (proactively); or the supplier responds, reacts, corrects, tries to recover, and suffers the consequences. Unfortunately, the latter approach may also cause the customer to suffer.

Be sure to document all exceptions (and their resolutions) identified during the review. Document all customer waivers that are granted. When the actual order or contract is received, it is crucial that it be closely compared to the tender (proposal) that was originally offered and with the exceptions/waivers. Sometimes what was thought to be understood and agreed upon has a mysterious way of changing in the final contract or order. Your customers may be notorious in this regard; so too may be your own marketing personnel—all the more reason for cross-functional reviews.

Don't forget to pay attention to all documents and data that form an official part of the contract language. They may include considerably more than just the customer's purchase order. The review process must identify and consider such items as purchase order supplements, contracts, drawings, specifications (customer and industry), regulatory standards and codes, electronic data, customer samples/models, referenced supplier or customer documents, and catalog descriptions. Furthermore, a change to any of these may necessitate additional review activity, although it will not be as extreme as the original effort.

You may receive a copy of a revised customer drawing or a modified spec with little or no mention of a contract change. Be careful! The consequences of such a little change may be very significant to the supplier and greatly affect the requirements to which the supplier is contractually bound. This raises an interesting point about changes to the contract.

The standard addresses amendments to a contract in paragraph 4.3.3. It requires that the supplier procedurally describe how an amendment to a contract is to be made, and how the change (and its effects) are transferred to all concerned (impacted/affected) functions within the supplier's organization. How this is done, to whom the change is transferred, and who the concerned functions are will depend upon the type and extent of the amendment. The process may not be as extensive or involve as many people/groups as the original review. Certainly, a change of quantity, delivery date, or carrier will not be handled the same as a change to technical specifications or quality requirements.

So who might be involved throughout the contract review and amendment process? Actually, to varying degrees, I've seen all of the following functions represented. Here's an open list from which to choose.

- Design
- R&D
- Engineering
- Packaging
- Shipping
- Purchasing
- Materials
- Inspection
- Quality
- Health/safety

- Marketing

- Manufacturing

- Production

- Process planning

- Accounting

When making the selection, remember the following points.

- Who is knowledgeable?

- Who is impacted/affected?

- Who is responsible for compliance?

- Who is authorized to decide?

- Who is interested?

- Who will provide resources?

The last clause under contract review is paragraph 4.3.4, Records. It simply says that records of contract reviews shall be maintained and refers to paragraph 4.16, Records. This reference provides somewhat of a definition. It says that quality records are to be maintained (identified, generated, collected, indexed, filed, stored, retained, dispositioned, and protected) in order to demonstrate (give objective evidence of) conformance to specified requirements (in this case, element 4.3 of the standard) and to the effective operation of the quality system. Notice that the standard is not talking about all records. Not all records are quality records.

Your system documentation (procedure, work instruction, records list) needs to explicitly identify which specific quality record(s) demonstrate contract review (and approval). It is not acceptable, or definitive, to say just *contract records*. That does not tell your contract reviewers, your internal auditors, or your registrar anything. Consider the following:

- What are the particular documents (name, number, and so on) or forms?

- Will they be signed, initialed, or stamped?

- Will they be electronic or hard copy, and where will they be found?

- What document/record captures evidence of the following?

—Review

—Raised exceptions

—Amendments

—Participation by all required reviewers

—Customer acceptance

—Approval

—Resolutions

—Waivers

I've seen some very clever adaptations of checklists for the contract review process. The checklist, though partially self-explanatory, is usually accompanied by a detailed work instruction. The process goes something like this.

The form is initiated by marketing, although it may have been preceeded by a request from a customer service field engineer. The marketing manager or her customer service rep indicates on the checklist the type of transaction this is.

Order type: _____ Repeat _____ New catalog _____ New special

Specification: _____ New _____ Revised

Drawing _____ New _____ Revised

Product modification _____

Process modification _____

Project work: _____ New _____ Extension

Developmental: _____

Price _____ Delivery _____ Special _____

A knowledgeable and authorized person (or team) working against predetermined and documented criteria checks off the functions that must review the contract, order, change, and so on.

Take a look at the following sample contract/order review form. It acts as a routing to all necessary people, signifies their review and eventual approval/rejection, and demonstrates final closure of the process. It accommodates recording of essential comments, exception notes, and pertinent attachments. Of course, many other pertinent details need to be recorded. Properly applying such a document provides excellent control and produces a very effective quality record.

Contract/Order Review						
Department	Reviewed	Information	Approved	Rejected	Comment #	Attachment #
Design	_____	_____	_____	_____	_____	_____
Inspection	_____	_____	_____	_____	_____	_____
Manufacturing	_____	_____	_____	_____	_____	_____
Materials	_____	_____	_____	_____	_____	_____
Planning	_____	_____	_____	_____	_____	_____
Engineering	_____	_____	_____	_____	_____	_____
Quality	_____	_____	_____	_____	_____	_____
R&D	_____	_____	_____	_____	_____	_____
Purchasing	_____	_____	_____	_____	_____	_____
Production	_____	_____	_____	_____	_____	_____
Marketing	_____	_____	_____	_____	_____	_____
Packaging	_____	_____	_____	_____	_____	_____

Comments:

1. _____

2. _____

3. _____

Approval after full review, resolution of exceptions, and final acceptance: _____ Date: _____

ISO 9000/ QS-9000 and the Plant Services Organization

From time to time I come across companies that are not fully convinced that the ISO 9000/QS-9000 quality systems standards really do govern and affect virtually their whole organization. Now, they have overcome the narrow-minded thinking this is just "quality's program." They have even come to appreciate the quality-related roles of many of the other major departments such as purchasing, process planning, engineering, and so on. But they wrestle with the notion that some of the peripheral departments can also play an important role.

They are reluctant to enlist these groups like plant services under the control of the quality system. This is for three basic reasons.

1. Management of the company and personnel in departments like plant services may not understand what the standards actually say and/or do not grasp the applicability of the requirements to their department.

2. These peripheral, support organizations may not be accustomed to the structure, disciplines, and controls normally associated with formal quality management systems.

3. They may not fully appreciate the contributions being made to the success of the quality system by these departments and do not appreciate their exposure and the consequences to the organization if these departments fail to perform properly.

Last year I was invited to speak to an audience at a plant engineering/plant services trade show in Charlotte, North Carolina. Fortunately, the attendees were willing to hear about ISO standards. Good thing, because I would have been hard-pressed to shuffle through my rendition of everything you wanted to know about aqueous ultrasonic washing systems. While I felt competent and was eager to discuss the standards and quality systems, I was challenged as to how to relate these subjects to a bunch of plant engineers and maintenance supervisors. Having visited 70 to 80 different manufacturing facilities in the previous four years, I was certain that these men and women were in fact involved with activities prescribed by the standards, and that they do indeed impact the compliance and effectiveness of the quality system.

My best approach to enlightening the audience seemed to focus on presenting real-life situations. I created a fictitious company, Big Boy Bearing Company, that was actually a composite of real companies—the three I've worked for during the past 25 years. Of course, I did add a few creative touches just to drive home key points.

This chapter takes you through eight examples at Big Boy Bearing Company. Each describes activities and responsibilities of various groups that comprise Big Boy's plant services organization. For each example, we'll also discuss requirements in the standards that govern or apply to those functions. You should be able to relate your company's plant engineering and plant services organization to some of Big Boy's and associate the pertinent aspects of the standard.

ISO 9000/QS-9000 requirements include (among others) the following:

- Contract review
- Control of purchased products and services
- Process control
- Control of inspection, measuring, and test equipment
- Handling, storage, packaging, preservation, and delivery
- Control of quality records

- Training
- Statistical techniques

A Case Study of Big Boy Bearing Company, Maker of Antifriction Ball and Roller Bearings

Big Boy's products are

- Machined and inspected to precision tolerances
- Handled with the utmost care
- Cleaned, preserved, and packaged to exacting specifications
- Stored in controlled environments until shipment

Big Boy's plant services organization consists of

- The facilities maintenance department
- The plant engineering department
- The equipment engineering and maintenance group
- The packaging systems group
- The environmental control group
- The electromechanical support group

How do the requirements of the standards apply to Big Boy's plant services organizations?

Example 1: Contract Review (4.3)

Several of the plant services supervisors participate in the review of contracts from Big Boy's customers, looking for requirements related to machine capability and product cleanliness, preservation, and packaging.

The standard requires the supplier to ensure that contract requirements are adequately defined and documented and that the supplier has the ability to meet those requirements *before* accepting the contract. Records of this review shall be kept.

Example 2: Purchasing (4.6)

Big Boy's plant engineering department uses outside subcontractors for the following functions:

- Machine repair and overhaul
- Inspection equipment calibration
- Maintenance of bearing cleaning systems
- Installation of environment control systems
- Fixture manufacturing and repair

The standard requires the supplier to evaluate, select, and control subcontractors. Purchasing documents shall clearly describe all requirements and shall be reviewed and approved before release.

Example 3: Process Control (4.9)

Big Boy's critical process equipment is maintained by the equipment maintenance group to ensure continued capability. The group also monitors and keeps records of key equipment/process characteristics. Equipment engineering ensures that qualification tests are performed before equipment is approved and released to production. Routine maintenance of the environmental systems is done by the environmental control group. The electromechanical support group provides technical expertise for computer numerical control machinery. Big Boy's preventive maintenance program is administered by plant engineering. Vital housekeeping is done by facilities maintenance.

The standard requires that the supplier ensure that processes affecting quality, including the following, are carried out under controlled conditions.

- Use of suitable equipment and working environment
- Monitoring of key characteristics
- Approval of equipment and processes
- Suitable maintenance to ensure equipment capability

Example 4: Control of Inspection, Measuring, and Test Equipment (4.11)

Some special inspection, measuring, and test equipment (IM&TE) is maintained and calibrated by subcontractors that are selected and controlled by equipment engineering. Regular recall and calibration of IM&TE is done through the equipment maintenance group's preventive maintenance and work order tracking system. A controlled environment in the calibration laboratory is assured through maintenance and monitoring by Big Boy's environmental control group. The electromechanical support group gives expertise on state-of-the-art video, laser, and computerized inspection equipment.

The standard requires the supplier to have procedures to control, calibrate, and maintain IM&TE, including use of suitable calibration environments and protection/storage of IM&TE.

Example 5: Handling, Storage, Packaging, Preservation, and Delivery (4.15)

Big Boy's packaging systems group installs and maintains the bearing cleaning, preservation, and packaging systems and is responsible for their performance. Equipment engineering and maintenance groups select, install, and support all automated product handling systems (conveyors, robots, and so on). Finished product storage elevators, racks, and storerooms are the responsibility of the facilities maintenance department. Temperature and humidity controls in store rooms is assured by the environmental control group.

The standard requires the supplier to have procedures for the handling, storage, packaging, preservation, and delivery of product.

Example 6: Control of Quality Records (4.16)

All of Big Boy's plant services groups generate and maintain a variety of vital quality records relating to the following:

- Calibration of IM&TE

- Process equipment qualification and approval

- Subcontracted service purchasing

- Maintenance technician training

- Equipment preventative maintenance

- Environmental control

- Product cleanliness audits

The standard requires the supplier to have procedures for the identification, collection, indexing, access, filing, storage, maintenance, and disposition of records that demonstrate the effective operation of the quality system.

Example 7: Training (4.18)

Supervisors in Big Boy's plant services organization conduct training for a variety of personnel that perform work affecting quality, such as the following:

- Calibration technicians

- Industrial engineers

- Equipment repairmen

- Electronic technicians

- Environment control engineers

All plant services employees are trained in practices, procedures, and work instructions.

The standard requires the supplier to have procedures to identify and provide for the training of all personnel whose activities affect quality.

Example 8: Statistical Techniques (4.20)

The equipment engineering group uses statistical techniques to determine machine capability during equipment run-offs prior to approving and release to production. Plant engineering and equipment maintenance groups apply statistical techniques to predict equipment wear and to establish preventive maintenance schedules. IM&TE intervals are determined by the equipment maintenance group using statistical techniques.

The standard requires the supplier to identify the need for statistical techniques and have procedures to implement and control their application.

Summary

The plant services organization at Big Boy Bearing Company is directly affected by several elements in the ISO 9000 and QS-9000 series standards. Plant services may be indirectly affected by other elements in the standards such as

- Management reviews
- Internal quality audits
- Corrective and preventive action
- Control of nonconforming product

Plant services activities definitely help Big Boy meet quality system requirements. Plant services is a key contributor to Big Boy's successful ISO 9000/QS-9000 registration.

Considerations for Selecting Your Third-Party Registrar

When I first became involved with the ISO 9000 industry back in 1990–1991, there were just a couple dozen registrars doing business in the United States. Today there are nearly 70. Last year I attended ASQC's 50th Annual Quality Congress in Chicago, and the Auto Show in Detroit. At both events, I was amazed at all the registrars with their wares on display. I had not heard of many of them previously. I refer to this incredible growth as the *mushroom syndrome*. Ever notice how, on a freshly mown lawn, mushrooms spring up over night, only to die after a couple of days in the hot sun? Well, the mushrooms are all over the front yard, but some have already withered and others are showing signs of exposure. There are many important characteristics of quality systems registrars that you must consider before making a responsible, informed selection. But four are essential. Above all else consider reputation, credibility, recognition, and stability. Today, more than ever, stability is becoming increasingly important, now that the proliferation has begun.

There are other attributes and characteristics that also deserve your attention. In this chapter I will raise your level of awareness and arm you with some fairly probing questions. With them you can query prospective registrars from a position of knowledge and authority. Teachers sometimes tell students, "Just ask me if you don't understand." But, when I took calculus (the first time), I was so ignorant that I couldn't ask even semi-intelligent questions. Doesn't it make sense that, of the 70 or so registrars, there should be differentiating qualities? There are. Not every doctor can graduate first in his or her class; someone has to be last, and many are in between. Registrars are similar. Wouldn't it be great to have a questionnaire? This chapter is one!

There is just one snag though. When you get the answers, will they be right or wrong? Will you be able to evaluate the responses? Will they mean anything to you? Let me explain with an example. I've been looking for a cellular phone for my wife. Not knowing anything about this technology, service agreements, roaming, or line charges, I went to the first dealer. After 30 minutes I was overwhelmed with facts. The salesperson had an answer for every question. Did I buy the phone on the spot? Of course not. This first episode was only my point of reference. From this benchmark I could intelligently evaluate the other dealers.

Throughout the remainder of this chapter I'll take a similar approach. I will raise meaningful questions that you should also ask the candidate registrar. Then I'll answer each question with hypothetical, desirable responses based on my experience. These answers will be the point of reference against which you can evaluate responses from the other candidates.

What things should you ask about when choosing a registrar? There are a number of principal attributes that must be evaluated. They include the registrar's

- Accreditations

- Global presence

- Assessment staff

- Costs and value

- Client base

- Stability and maturity

- Regional support

- Own quality system

- Service offerings

- Interpretations and applications

Question: Ask about accreditations.

- Are your accreditations direct or indirect?
- How many accreditations do you have?
- What type of accreditations do you have?
- Are you expanding your accreditations?

Answer: Desirable responses include the following:

- All accreditations are direct.
- There are no memorandums of understanding (MOUs).
- We currently have 10 or more accreditations.
- We are continually seeking more accreditations.

Question: Ask about stability, maturity, and presence.

- How many years have you been in business?
- What is your growth rate and expansion plan?
- How many personnel do you employ?
- Are you national, international, or global?
- Do you have established offices?
- Are your assessors available internationally?

Answer: Desirable responses include the following:

- Profile of parent company
  - Founded in the early 1800s
  - Global operations
  - Several thousand employees
  - Staffed offices in more than 130 countries
  - Present on several continents
  - Started in the shipping industry
  - Now in many major verification businesses

- Profile of quality system registrar unit
  - —Founded in late 1980s
  - —Global operations
  - —Several hundred assessors (employees)
  - —Staffed offices in dozens of countries
  - —Present on many continents
- Profile of North American operations
  - —Founded in 1990
  - —Headquartered in major city (for example, New York)
  - —More than 50 personnel (direct employees)
  - —Many regional, staffed offices throughout the United States

Question: Ask about regional support.

- Do you have local or regional offices throughout the United States?
- Do your offices have dedicated client accounts?
- Are there qualified assessors in each region?
- Are services customized and personalized to clients?

Answer: Desirable responses include the following:

- Eight fully staffed offices
- Each has 100 to 300 accounts
- Dedicated assessors for regions
- Executive directors, staff, and assistants
- Expanding in North America (Mexico, Canada, Caribbean islands)

Question: Ask about the registrar's quality system.

- Has it been independently assessed?
- Does it meet European Norms EN 45011/EN 45012 for registrars?

- Does it meet international CASCO Guides 61/62 for registrars?

- Does it meet the ISO 9001 standard?

- Is it standardized worldwide?

- Is it fully and formally documented?

- Is it fully and directly accredited?

- Does it comply with the QS-9000 code of practice?

Answer: Desirable response include the following:

- Candidate should respond affirmatively to all these questions.

Question: Ask about the registrar's services.

- Do you provide ISO 9000 registrations?

- Do you provide QS-9000 registrations?

- Do you perform system pre-assessments?

- Do you offer TickIT (software) registrations?

- Do you offer CE Marking support?

- Are you a notified body (competent certification body, notified by the member states of the European Community as being in conformity to EN 45000)?

- Do you provide ISO 14000 series (environmental) registrations?

- Do you maintain a small firms scheme?

- Can you provide on-site assessment planning?

- Do your accreditations cover pertinent SIC/EAC codes?

Answer: Desirable response include the following:

- Candidate registrar should respond affirmatively to all of these questions.

Question: Ask the about the registrar's assessment staff.

- What is the maturity level of assessors?
- What type and amount of training do assessors get?
- What is the experience and background of assessors?
- Do you use subcontractors, and how much?
- How are assessment personnel screened and qualified?
- What credentials do the assessors have?

Answer: Desirable responses include the following:

- Profile of typical assessor
  - —More than 20 years professional experience
  - —More than 15 years in senior technical positions
  - —More than five years in managerial positions
  - —Fifty percent are registered lead assessors
  - —One or more professional credentials
  - —Certified by IQA/RAB; ASQC CQA certification
  - —Nearly all assessors are full-time employees
  - —Limited subcontractor usage on exception
  - —Eight to 10 on-site centralized training days yearly
  - —Extensive/thorough screening before hire
  - —Qualified for QS-9000, TickIT, and other specialty schemes
  - —Several interviews and observation audits before hiring
  - —Multiple performance reviews throughout year
  - —Technical scope assured through validation process

Question: Ask about the registrar's client base.

- Do you have well-known clients?
- Do clients represent diverse company sizes?
- Are clients from multiple business sectors?

- How many clients do you have?
- What is the growth rate of your client base?
- How many certificates have you issued worldwide?

Answer: Desirable responses include the following:

- A few Fortune 100 companies
- Many Fortune 500 companies
- Very familiar with small companies as well
- Companies with as few as one to five employees
- Companies with as many as 4000 or more employees
- Clients in service and manufacturing industry
- Clients in distribution and R&D
- Multinational clients
- Multiplant companies
- Broad range of industry sectors including chemical, machining, electronics, communications transportation, plastics, automotive, computers, textiles, software, distribution, and metal fabrication
- Many thousands of certificates issued to date

Question: Ask about costs and value.

- What are the included costs?
- What are the add-on costs?
- What are the man-day labor rates?
- Are the quotations straightforward and simple?
- How are the number of assessment man-days determined?
- Specifically, what are the costs for
  —Travel time
  —Quality manual reviews
  —Report writing

—Clearing nonconformities

—Holiday and weekend travel time

—Application fees

—Record maintenance fees

—Travel cost mark-ups

Answer: Desirable responses include the following:

- Man-days meet guidelines of accreditation body
- Man-days meet guidelines of automotive sector
- Quotations should be inclusive of all money
- There should be no add-ons or hidden costs
- Expect additional charges only for

  —Special follow-up or revisits

  —Visits for scope extensions

  —Actual travel expenses at cost

  —Certificates and extra originals

- Quotations may include

  —Discounts for schedule attainments

  —Special rates for small firms

- There should be no extra charges for

  —Quality manual reviews

  —Travel time

  —Necessary holiday/weekend travel time

  —Necessary extended work days

  —Mandatory record maintenance

  —Writing of reports

  —Mandatory auditing of off-shifts

  —Clearing of nonconformance documents

  —Application costs

—Postage, faxes, copies

—Schedule and agenda preparation

—Travel accommodation arrangement

Question: Ask about interpretations and applications.

- Are they based upon real experience?
- Are they objective and not arbitrary?
- Are they consistent over time and among assessors?
- Are they reasonable for your industry and products?
- Are they based on good business sense?
- Are they nonprescriptive?
- Is interpretive assistance available?

Answer: Desirable responses include the following:

- Proven, mature professionals
- Assessors have hands-on experience
- Ten percent theory and 90 percent practical application
- Must add value and not burden
- Assessors understand the intent of the standard
- Regular assessor team mixing for consistency
- Assessors have each performed 80 to 120 audits
- Experts at applying to your products, processes, and business

Question: Ask about miscellaneous items.

- What is the duration and man-days for each event?
- How long is the certification period?
- How can the certificate be renewed?
- Can the scope of the registration be expanded and how?
- What is the process for clearing nonconformances?

- Can additional certificates be obtained without re-assessment?
- What is the format and style of pre-assessments?
- When is it best to select the registrar?

Answer: Desirable responses include the following:

- Registration man-days are a function of the standards and other audit criteria.
- Man-days are a function of company size and process complexity.
- Man-days may be a function of scope extensions.
- Pre-assessments take approximately 75 percent of the registration man-days.
- Pre-assessments are performed like a dress rehearsal.
- Pre-assessments can be customized by time or focus.
- Surveillance audits take approximately 25 percent of the registration man-days.
- Renewal assessments take approximately 25 to 50 percent of the registration man-days.
- Three-year certification periods.
- Minimal evaluations needed to renew certification.
- Scope extensions granted with little or no assessment.
- Most nonconformities cleared on-site by fax, phone, or mail.
- Only major nonconformances require a revisit to clear.
- Future certificates are provided without additional assessment.
- Choose a registrar early for help with interpretations, scheduling, scope selection, on-site planning, and team assignment.

10 An Effective Document and Data Control System

The element of the ISO 9000 series standard that is most often cited as being nonconforming is control of documents and data. Having personally assessed 50 or 60 document and data control systems, I've seen just about every way possible to skin this old cat. And I understand why this element is so often a problem. There are several contributing factors.

1. People think in terms of controlling only the conventional types of documents such as the quality manual, quality procedures, specifications, and drawings.

2. This element affects virtually every department throughout the entire organization, and most companies have hundreds of documents and data in dozens of different formats and media, so there are lots of opportunities for failure.

3. Responsibility for control is often delegated to several different people and groups, each of which does things a little (or a lot) differently.

4. The control system is not well defined and understood. It is overly complicated, and therefore cannot be effectively and conve-

niently assessed by the internal auditors, resulting in hidden accidents waiting to happen.

5. The requirement to control data is overlooked, ignored, or poorly addressed, most likely because people may not appreciate the impact that uncontrolled data can have on the system or because the control of data presents a few new and unpleasant challenges.

6. The system has failed to identify the various types of documents and data that are subject to this requirement or, if it does, people are overwhelmed by the results.

7. Creators of the system are mad scientists that try to develop a one-size-fits-all approach. They think that every kind of document and every medium of data must be controlled in the same way. This can only result in an out-of-control Frankenstein monster.

You may be surprised at the different types of data media being used in your company. Here are a few kinds.

- CD-ROMs
- Microfilm
- Personal computer hard drives
- Computer numerical control tapes
- Magnetic tape
- Computer diskettes
- Mainframe files
- Printed paper
- Microfiche
- Imbedded software
- Hard copy drawings

There are a number of issues that need to be considered by your document and data control system.

Scope of the Requirement

The standard sets the scope of applicability for this requirement to include all documents and data that relate to the requirements of this standard and to documents of external origin (like customer drawings and specs, industry and national standards, regulatory codes, and so on).

That's a fairly tall order. Any internally produced document and any internally produced data that are pertinent to any activity prescribed by the standard and/or your own quality system are subject to an appropriate type and level of control. The same can be said of externally produced documents and data pertinent to your operation/system. The good news is that the standard does not mandate the type and extent of control (provided all basic requirements are satisfied), nor does it require that all documents and data be controlled in the same way.

Thoughts on Implementation

The quality system procedure needs to substantiate and logically justify *not* controlling other types of documents and data. What are the criteria for establishing control, or not? Have a means for distinguishing controlled documents from those that are not—for example, colored paper, colored borders/headers/logos, controlled when in red stamps, uncontrolled or reference only stamps.

Define guidelines for permissible use of uncontrolled or reference only or courtesy copies of documents and data. Identify what documents and data are under control. For example,

- Quality assurance manuals
- CAD/CAM files
- Tabulated data sheets
- Internal audit checklists
- Calibration instructions
- Quality plans
- Drawings
- Test data files

- Computer data files
- Machine preventive maintenance plans
- Engineering specifications/standards
- Design/fabrication/inspection software
- Quality system procedures
- Forms
- Standards and codes (external)
- Flowcharts/diagrams
- Process control documents
- Customers' documents
- Administrative procedures
- Manufacturing work instructions

Define guidelines for permissible use of uncontrolled or reference only documents/data or courtesy copies. Describe the basic/core attributes of controlled documents (as applicable). For example,

- Controlled paper/stamps
- Revision indicators
- Scope
- Approval signatures
- Pagination
- Ownership
- Primary/secondary/responsibilities
- Responsible and affected organizations
- Prescribed formats, content, headers, and so on
- Unique identifiers (name/number, and so on)
- Author/originator/approver
- Nature of the change (not just location)

Describe unique naming/number schemes. For example,

- Quality policy documents   "QP . . ."
- Quality system procedures   "QSP . . ."
- Quality work instructions   "QWI . . ."
- Drawing number   126729.1c (component 1, rev. 2)
- Formulation   AD27 (family, product, type, recipe number)
- Computer file   A:———.123

Address who assigns and how document and data indentifiers are issued to prevent redundancies. Describe various levels of control, which may be different. For example,

- Quality manual versus group level W/I
- Level of approval
- Extent of distribution
- Frequency of changes
- Redline options (handwritten, authorized modifications in red)
- Secondary distribution schemes
- Wall postings

Ensure that changes are communicated. Consider the following:

- Distribution cover letters
- Nature of changes (within/attached)
- Revision level identifier (number, letter, date, and so on)
- Document training

Identify specific, pertinent quality records that demonstrate effective operation of the document control system. For example,

- Internal/external transmittal forms
- Change request/initiation forms

- Change approval sheets
- Distribution/revision notices
- Signed receipt forms
- Training records
- Revision histories
- Historical copies

Maintain a master list(s) or equivalent control procedure showing the following:

- Document/data identifier
- Document/data revision indicator

Identify authorized personnel for review and/or approval of controlled documents/data. Consider the following:

- Usually only position title.
- Reviewer may be different than approver.
- Document control coordinator/custodian.
- Who has responsibility to secure approvals?
- Authority/responsibility may vary with nature/type of document or data and with degree of control required.
- Changes different than original approved.

Address the protection of controlled documents/data. Consider the following:

- Secured originals
- Password deactiviation
- Read-only versus edit options
- Use of white-out
- Electronic backup
- Password protection
- Controlled signature entry field

- Control coordinator
- Initialed/dated modifications

Describe the method of distribution/recall.

- Mailed versus hand-carried versus electronic
- Acknowledgments/receipts
- Passive versus assertive system
- Secondary/lower-tier systems
- Internal versus external distribution/recall
- Return versus destroying of obsolete versions
- Customized versus standard distribution
- Distribution memos

Food for thought: consider the following:

- Is the system auditable?
- Can its suitability and effectiveness be demonstrated?
- Centralized versus decentralized document/data control
- Formal change request system
- Locally controlled documents

11 Knowing Value Versus Price: What Do You Really Get?

It has been said that some people know the price of everything and the value of nothing. I'm a testimony to that—at least when I was young. During our early years of marriage, money was tight. Being a lover of home/appliance/auto repair (not only my wife), I was always buying tools. Our local lumberyard had the practice (trap) of displaying shiny tools at discount prices, in a special bin near the checkout register. A wonderful assortment of handheld devices at cheap prices, their quality was low and reflected poor value. But they were cheap so I bought them, and bought them again several times over. Before too many years passed, I learned I had spent more than if I'd purchased good quality, high-value tools at a fair price.

My motto now is to buy the very best quality tool I can possibly afford. Now my tools rarely have to be replaced, unless of course, one of them sprouts legs and walks off—as my son asserts they sometimes do. Though it took more than 20 years to learn this lesson, I can now apply it to nearly everything I purchase.

Permit me to offer my own definition of value. *Value* is a derived benefit expressed as a function of quality and price. The greatest value is derived from the highest quality (and/or quantity) at the lowest price. For the math lover in you, the value quotient is $V = Q/P$.

If the price and quality/quantity both vary proportionately (up or down), the value remains the same. The only ways to realize greater value are to do one or both of the following:

1. Reduce the price P while maintaining Q (or increase Q)

2. Increase the quality/quantity Q while maintaining P (or reduce P)

There is absolutely no way to assess value without knowing Q. Price P alone won't tell you a thing about value. Price can only indicate if the product or service is cheap.

The book you are reading represents true value (my publisher loves me). It was sold at a reasonable price and contains loads (quantity) of great (quality) stuff. What a smart shopper you are! You may be shopping for a registrar for your quality system. If so, here's an opportunity to apply the value quotient formula. Obviously, the registrar providing the greatest value should be given first consideration. Note that I did not say the registrar with the cheapest price.

While the equation itself is simplistic, determining its factors (Q and P) is not so simple. Can you believe that some registrars may try to disguise some of their price? If they succeed, then their perceived value is falsely inflated—at least in your eyes. Did you ever purchase a car from a dealer and trade in your old "only driven on Sundays" cream puff? Chances are you never knew the true price of the replacement car or what you really got for "old faithful." Why? Because the salesperson disguised the prices so you would perceive greater value in the deal. You can tell that I've bought my share of cars, too.

Back to selecting your registrar. Registrars are in business to make money—even the ones that are reported to be nonprofit. That's okay. They should and must in order to stay in business. You definitely want your registrar to be there for you, year after year. As with tools and cars, some registrars deliver more value than others. Remember, price alone tells you nothing about value: $V = Q/P$.

Your task is to determine Q and P. The dollar amount readily apparent on the registrar's quotation is probably not the complete price (P). I can say with confidence and from firsthand experience that quotations can and should be simple and straightforward, and the stated three-year contract costs should be nearly all-inclusive. With some other registrars you may have to scratch and dig, as they may not like to call attention to some of their charges until after you sign the contract. The most forthright registrars will gladly share the requested information with you.

Following is a list of nearly every cost you can potentially incur with any registrar. I have taken the liberty to partially complete the form using an imaginery example (ACME-ISO Ltd.) so that you'll have a point of reference against which to compare other candidate registrars.

Price (*P*) Components for Registrar ACME-ISO Ltd.			
Component	**Y/N**	**Rate**	**Cost $**
• Pre-assessment audit	Y	$900/man-day (4)	$3600
• Registration audit	Y	$1200/man-day (6)	$7200
• Maintenance audit	Y	$1100/man-day (8)	$8800
• Special follow-up visit	Y	$1450/man-day (2)	$2900
• Routine travel time	Y	1 man-day per event (8)	$7200
• Weekend travel time	Y	$60/hour (est. 4)	$240
• Holiday travel time	Y	$75/hour (est. 4)	$300
• Quality manual review	Y	$375 each (twice)	$750
• Audit reports	Y	$115 each (est. 8)	$920
• Travel markups	Y	actual + 4.0 percent (est.)	$180
• Off-shift coverage	Y	$125/hour (est. 4)	$500
• Extended workdays	Y	$125/hour (4)	$500
• Application fees	Y	$175	$175
• Record maintenance fees	Y	$800/3 yr.	$800
• Certification maintenance fee	Y	$1000/3 yr.	$1000
• Nonconformance clearance	Y	$50 each (est. 18)	$900
• Standard postage	Y	actual + 25 percent (est.)	$50
• Phone/fax	Y	actual + 25 percent (est.)	$50
• Original certificates	Y	$900 each (3)	$2700
• Duplicate certificates	Y	$50 each (3)	$150
• Administrative fees	Y	5 percent of contract (est.)	$2200
• Information calls	Y	$45/hour (est. 2)	$90
• Off-site planning	Y	$60/hour (est. 4)	$120
		Estimated true price (*P*)*	**$ 41,325**

*Estimate based on total three-year contract costs for a company with 250 people (first shift) seeking ISO 9001 certification.

The remaining component of the value quotient formula is Q (quality/quantity). Q, like P, is a composite of many factors. The numeric you assign to each Q factor will depend on your particular company and quality system. Those factors that are considered more important will carry a higher numeric. Just be sure to consistently apply them to all registrars. May I suggest that you assign numbers (0 to 5) for each Q factor—depending on its importance and the registrar's ability to deliver, with zero for none and five being the highest.

On page 115 is a list of many Q factors that you may wish to consider. They refer to characteristics/attributes of registrars. For practical purposes, you may choose to reduce or consolidate it. Again, I've taken the liberty to partially complete the form using the same imaginery registrar (ACME-ISO Ltd.) as an example so that you'll have a point of reference against which to compare other candidate registrars.

Now that you have determined the true price (P), and have summed the total quality/quantity (Q) score, you are ready to compute the final value quotient (V) using the formula $V = Q/P$ for each registrar. I've completed the following summary form with some hypothetical numbers.

Value Quotient Summary Sheet

Registrar	Q Score	/	Price (P)	=	Value (V)
AAAAAA	110	/	$17.7K	=	6.2
ACME-ISO	144	/	16.5	=	8.7*
CCCCCC	89	/	15.1	=	5.9
DDDDDD	137	/	20.3	=	6.7
EEEEEE	96	/	13.5	=	7.1

*Highest value

My family used to live in upstate New York, where a retail men's clothier had the slogan, "The informed consumer is our best customer," or something like that. Take the time to complete these exercises, finding the real value of each registrar. You'll find the task to be a real eye-opener. It will take away a lot of subjectivity and mumbo jumbo, and your management will appreciate that you've done your homework. You'll like having something to hang your hat on. The relationship between your company and its registrar will hopefully last for many years. Pick the registrar based on value, not just price.

Quality/Quantity (Q) Components Registrar ACME-ISO Ltd.

Component	Y/N	Q Factor
• Employees versus subcontractors	Y	5
• Caliber of assessors	Y	5
• Credibility/reputation	Y	5
• Recognition	Y	5
• Global presence	Y	5
• Regional offices	Y	5
• Directly accredited	Y	5
• Operates without MOU	Y	5
• Core business is registration	Y	5
• Assessors multicertified	Y	4
• Scope authentication	Y	5
• Assessor prehire qualify	Y	5
• Optional pre-assessments	Y	5
• Customized pre-assessments	Y	5
• Industry experience	Y	5
• Diverse customer base	Y	5
• Full-service provider	Y	5
• Stability/years in service	Y	5
• Variety of customer size	Y	4
• Office responsiveness	Y	5
• Assessor technical knowledge	Y	4
• Offer TickIT for software	Y	5
• Offer environmental ISO 14000	Y	5
• Offer CE Mark support	Y	4
• Registrar is a notified body	Y	4
• Offer automotive QS-9000	Y	5
• Rapport with office staff	Y	5
• Most widely accredited	Y	5
• Tri-annual assessor training	Y	5
• Dedicated client accounts	Y	4
Composite quality/quantity (Q) score		**144**

Note: Use zero for "No," and 1 to 5 for "Yes" responses

CHAPTER

12 The Quality Assurance Manual Adequacy Audit

As described in chapter 2, the quality assurance manual (quality policy manual, quality control manual) represents the top-level system documentation. It is subject to an adequacy audit by the registrar, usually prior to the on-site assessment. The ISO 9000 standards actually have very little to say about the manual. In paragraph 4.2.1, Quality System—General, the standard says that the supplier must prepare a quality manual that covers the requirements of the standard, that includes or makes reference to the quality system procedures, and that outlines the structure of the documentation used in the system. ISO 10013 is a great reference and guidance tool.

The nature and content of the review will vary depending on the registrar and the individual actually doing the audit. Usually, the lead assessor assigned to perform your company's system assessment evaluates the manual. This is the preferred approach. However, some registrars will use office personnel dedicated to doing manual reviews for all of their clients. In the latter case, the results of the audit, and the manual itself, are forwarded to the lead assessor in order to maintain some continuity.

While expectations and required level of detail will also vary among registrars, there are quite a number of minimum requirements that nearly all accredited registrars will look for. You, as the customer, should also have expectations of the registrar's audit and subsequent report.

In alphabetical order, let's first identify the minimum checklist items that the registrar will expect of your manual.

Approval: Does the quality manual bear evidence of approval (for example, name, position, signature, date) by duly authorized and responsible parties?

Clarity: Has the manual clearly described the quality system and is it definitive and decisive?

Clauses: Are all clauses of the standard addressed in the manual and is an explanation given for those that are not applicable?

Consistency: Is the manual consistent in level of detail, style, and format throughout the document?

Contents: Is a table of contents or index provided as appropriate?

Controlled: Is the manual an official, controlled document?

Cross reference: If the manual is not directly aligned and numbered with the standard, does it contain a cross-reference?

Definition: Are all unusual or company-specific terms, abbreviations, acronyms, and so on fully defined in the text or glossary as appropriate?

Detail: Are company commitments to each of the elements in the standard discussed in sufficient detail? Is the specificity such as to provide adequate overview of each system activity?

Elements: Are all elements of the standard addressed in the manual and explanation given for those that do not apply?

Factored items: Are registrar-specific requirements (such as factored items) addressed, defined, and applicability stated?

History: Does the manual include a revision history, change log, or equivalent means of showing the history of its change activity, describing the nature of changes to the document?

Identification: Do all pages of the manual bear company identification (such as name, logo, and so on)?

Management rep: Has the ISO management representative been identified in the quality manual or in another official document that is referenced in the manual?

Naming: Has the naming and numbering convention of the quality system documentation been identified in the manual?

Nature of changes: See **History**.

Organization: Is the company's management structure described in the manual and supported with an organizational chart as appropriate?

Pagination: Are all pages numbered and associated with others (page 1 of 38, page 2 of 38, page 38 of 38)? Are all charts, appendixes, tables, attachments, and so on adequately identified and/or numbered?

Personalized: Is the manual personalized and customized to your company and reflective of your quality system, not just a generic, off-the-shelf mimic of the standard?

Policy: Is the company quality policy described (stated) in the manual?

Procedures: Are all pertinent quality system procedures included within, referenced in (name, number, and so forth), or otherwise officially linked to the manual?

Responsibilities: Are all key responsibilities and authorities for activities specified by the standard fully identified in the manual?

Review: Does the manual prescribe its periodic review and contain evidence this is being done?

Revision: Is the revision level (number, letter, date, and so on) of the document and its individual component parts (pages, sections) clearly identified?

Scope: Does the manual fully describe the scope of the quality system?

Standards and codes: Are pertinent standards and codes that form a part of the quality system (military standards, industry standards, regulatory codes, and so on) identified in the manual?

Structure: Does the manual define the structure, or architecture, of the documentation used in the system?

Now let's consider some things that you should expect from the registrar. In no particular order of importance, they are as follows:

- A fairly detailed (but not full of minutia), official, typed report with an accompanying checklist

- A personal critique of the manual by the lead assessor or another authorized agent of the registrar

- An overall evaluation (overview) of the suitability and effectiveness of the quality manual as the highest governing document

- Identification of the strengths and weaknesses of the manual

- Clear identification and explanation of potential nonconformances or omissions that could complicate the on-site assessment

- A positive, upbeat, constructive evaluation of the document

- Reference to the official criteria used in the review (for example, ISO 9001, ISO 10013, registrar's procedures, and so on)

- Reinforcement and complimentary remarks for the good points

- Clarification and definition of unfamiliar terms used by the particular registrar (for example, factored items)

- An objective, nonjudgmental, decisive and definitive appraisal

- Nonprescriptive and nonconsultantory commentary

- A disclaimer by the document reviewer that a complete review/analysis of the entire system documentation will depend on further review of lower-level documents (procedures and so on) while on-site

- A description as to who, what, when, why, where (but probably not how as this may approach consulting) to address noted deficiencies, omissions, and nonconformances

- A critique of the manual as a systemwide document, not minor flaws

- A genuine offer by the reviewer to further discuss and clarify the results of the audit with you

- A reflection on each of the "shalls" specified in the main elements of the standard

- An official, well-thought-out report; professionally prepared, with an introduction, body, and conclusion/summary

- Polite, respectful, and courteous constructive criticism

- A conclusion as to the adequacy of the manual and mention of its probable success during the on-site system registration assessment

Only the most extreme problems with the quality manual should be sufficient cause to cancel, or even postpone, the on-site assessment. Virtually every issue can be resolved before or during the assessor visit. Timely submission of the manual to the lead assessor is important. Registrars seek a four- to eight-week lead time. Conditional upon your submission date, you should expect documented results of the adequacy audit two to four weeks before the assessor visit. This should provide ample time to make improvements and corrections and to cascade effects down to lower-level procedures if needed.

13 Corrective Action Responses to the Registrar

After many months (or one to two years) of planning, preparation, documentation, and implementation, followed by iterations of rigorous internal audits and probing management reviews, culminating with the third-party assessment and a dozen or so resulting nonconformities, it all comes down to this: How do you provide corrective action responses to the registrar in a way that will maximize their chances of being accepted and cleared?

Most accredited quality system registrars (for ISO 9000, ISO 14000, and so on), and all those that are accredited to register companies to QS-9000, require that all quality system nonconformities be cleared before the system can be registered and certificates can be awarded. Clearance is interpreted as complete resolution, including the following:

- Determining the system/root or isolated cause

- Identifying the corrective action(s)

- Implementing the corrective action(s)

- Providing objective evidence that the actions were actually taken

- Providing objective evidence of necessary/resultant training

- Providing evidence that the action was effective in correcting the problem

- Providing documentation of these aspects to the registrar

You would think that if a company makes it this far along on the road to registration, responding to a few measly nonconformances would be a no-brainer. You would think so, but it's not so!

After seeing 100 or so cases of companies fumble the ball on the three-yard line (holding up nonconformance clearance and the company's certificate, regurgitating paperwork, tying up the U.S. mail and fax machines, and tracking down lead assessors to have countless hours of phone discussions), it finally dawned on me that I should prepare a guide for corrective action responses. If the company would read and then follow the guidelines, it would certainly make everyone's life easier, especially the lead assessor's, whose responsibility it is to ultimately close the nonconformance documentation. This must be closed and validated so that certificates can be printed. Since the lead assessor is nearly the last link in the chain, guess who gets all the heat when the client is jumping up and down for its certificate? Once the lead assessor has the ball, he or she is the hold-up, regardless of the causes for all previous delays. Now, being associated with the registrar, I really do understand the value (and peace) that comes from a happy, certificated customer. From experience, I also understand that the vast majority of lead assessors are ready and willing to accept a reasonable and well-documented corrective action. However, nine times out of 10, the customer ties the assessor's hands by offering unacceptable information on one or more nonconformances.

What's all the fuss about acceptable corrective action responses (let alone perfect ones)? I'll explain. The best registrars have a dedicated certification manager whose job is to make sure that all paperwork, forms, and documentation are squeaky clean before submitting them to the national accreditation body. That body, in turn, authorizes the printing and award of the certificate to the assessed company. The really primo registrars have demonstrated such discipline and control that their certification manager is authorized to produce certificates directly, without submitting the entire documentation package to the accreditation body. This cuts the certificate award process time from two to three months to just two to three weeks. This advantage has helped us to get many certificates into Santa's sleigh and under the Christmas tree in the nick of time.

What's more, the documentation package is subject to scrutiny by the accreditation body when it performs an audit on the registrar's quality records. Yes, this really happens, and often! There is a lot riding on the accuracy and completeness of the paperwork, a crucial component of which is the nonconformity clearance documentation. So the clearance process has to be executed well—starting with the client. Recommended guidelines, enter stage left. Eureka, they worked!

Let's agree on a term to describe the official nonconformity document that the registrar uses. Each registrar calls it by a different name or number. For discussion's sake, let's call it the quality system nonconformity (QSN). The guideline checkpoints fall into four categories: corrective action descriptions, corrective action implementation, objective evidence, and attachments.

Corrective Action Descriptions

• Address each specific deficiency and example that is noted on the original QSN form. State the remedial action that was taken to correct each cited item.

• Address what global action was taken to purge the system/facility of other examples of the same, or similar, deficiency that may exist elsewhere, beyond those few examples listed on the QSN.

• If applicable, state the action taken to correct the underlying root systemic cause—the quality system weakness/omission/deficiency, or lack of compliance and discipline, that allowed the problem to occur in the first place.

• State whether or not localized and/or systemwide training has been taken to implement or reinforce the corrective measures.

• Identify any pertinent documents (procedures, work instructions, forms, and so on) that were created or revised in order to implement or reinforce the corrective measures. List their names, numbers, and latest revision level.

• Descriptions should be clear and concise. If they are too lengthy to fit in the allowed space on the QSN, the essence should be described in the response block, with elaboration given on a continuation page or on the back of the QSN form. Avoid simply saying "see attached" in the response block. Give clear indication of any and all attachments and continuations.

• Ensure that all mention of actions/activity are in the past tense. Use "were," "have been," "was," and so on—not "will be," "will do," "plan to," and so on.

Corrective Action Implementation

• State in past tense the corrective description. The action must have already been implemented before submission to the registrar. The response should not describe future or proposed activity—unless such activity is actually above and beyond that which is essential to resolving the specific deficiency.

• If the QSN form has a block for "date of completion," it must be a past date, not future.

• The ISO management representative, or other duly authorized party, must verify and attest that the corrective action, as described, has actually been fully and effectively implemented and should sign and date the QSN form.

Objective Evidence

• Provide supporting documentation to demonstrate that the described corrective action has been fully and effectively implemented. The better this is done, the greater the odds are that the lead assessor will be able to accept the response and clear the nonconformance either through mail or via fax—without an on-site revisit.

• Supporting documents should be official and may include the following:

—Changes to the quality manual

—New or revised quality system procedures

—New or revised work instructions or forms

—Statistical process control (SPC) charts

—Inspection and test records

—Calibration reports and certifications

—Preventive maintenance schedules and logs

—Management review meeting minutes

—Design review meeting minutes

—Design project plans

—Design validation results

—Training documents and attendance records

—New or revised job descriptions

—Internal audit schedules and reports

—Various quality records

—Reviewed and approved purchasing documents

—Photos, graphs, charts, and printouts.

Attachments

- Attachments should be

    —Uncontrolled photocopies (not originals) of official, approved, and controlled documents

    —Highlighted and/or underlined to clearly identify the changes pertinent to the corrective action response

    —Specifically referenced and identified (name, document number, revision) within the corrective action description block

    —Stapled or positively affixed to the original QSN form

- For small documents, include all pages. For larger documents (greater than 10 pages), include only those pages that show the document identification, nature of change, revision level, approval, and pertinent additions and/or changes.

- For multiple QSN nonconformity forms from the registrar, be sure that each one includes all pertinent attachments so that it becomes a complete, stand-alone package independent of other QSNs.

- Submit the completed original QSN form and all attachments to the lead assessor or designated office of the registrar. This should be done approximately 20 to 30 days before the end of the allowable clearance period to ensure timely closure by the assessor.

- Keep copies of everything!

Remember, lack of urgency on your part will not necessarily result in a state of emergency for the registrar or its lead assessor. Procrastinators beware!

There is another very special scenario that comes along once every 300 to 400 system nonconformances. Here's the pitch. The company has a nonconformance that it knows how to thoroughly and effectively correct. The catch is that the implementation time will (and must) take far longer than the clearance period allowed by the registrar (typically 90 calendar days). When this is genuinely the case (and not just an excuse for foot-dragging, preoccupation with meeting monthly production quotas, or unwillingness to commit necessary resources), the company should take three giant steps backwards, say "Simon may I," and then refocus on the problem. It should consider a two- or three- step (phase) approach to the solution. Possibly its original corrective action plan is the ideal, long-term, continuous improvement solution that should eventually be realized. If it's the right thing to do, then the company must commit to it and get started. Let's call this Phase II.

For now, perhaps there are some additional measures that can be taken immediately. When these are complemented and reinforced with some temporary (albeit costly, time-consuming, resource-intensive) actions, the combination will effectively treat (not cure) the system illness. Let's call this Phase I. There may even be intermediate phases (Ia, Ib, Ic) that unfold while Phase II is being developed.

Phase I (and any intermediate phases) must be effective (not necessarily efficient) in resolving the noncompliance—so much so that it could remain in place indefinitely, provided the company could tolerate the pain of the added cost, people, and so on (all are just band-aids). And it must be fully in place within the allowable clearance period.

Both Phase I and Phase II (and any intermediate phases) must be described in a documented plan and submitted to the registrar for acceptance. Based on this proposal and objective evidence that Phase I is working effectively, the lead assessor may be able to clear the nonconformance.

Then, in due time, after Phase II is ready to begin (possibly even running parallel to Phase I for a while), Phase I controls can be pulled out (band-aids removed). Phase II becomes permanent and may be verified as effective by the registrar at future visit.

Want a real-life example? A company failed miserably at internal calibration and control of IM&TE. This condition was justifiably written up and classified as a major nonconformity by the lead assessor. A few weeks later, the company responded with a corrective action plan that included hiring personnel knowledgeable of calibration systems; purchasing and installing a new computer and software for the gage tracking, recall, and calibration system; identifying, labeling and cataloging all IM&TE; and writing equipment-specific calibration procedures and work instructions. What a great plan! This was the company's Phase II and would take six to eight months for full implementation. But the registrar required clearance in 90 days. The company's registration and ISO 9001 certificate was in jeopardy. Corporate mandates were breathing down managers' necks.

Enter the lead assessor in shining armor, stage right. "Isn't there anything the company can do short term to properly calibrate its IM&TE before use?" I asked knowingly.

"Well, maybe we could corral all the IM&TE and only issue it out after it is calibrated by an outside service."

"That's good, but what about holding up production?"

"We could even hire a temporary person to calibrate in-house the critically needed stuff."

"Now you're cooking. But what about calibration records until the computer and software comes in, four months from now?"

"Easy, we have a clerk that can keep a manual card file till then."

"Sounds like a good Phase I to me. Go for it," I said.

Both Phases I and II were documented on the QSN form and continuation page. Before the end of the five-day assessment, the company had Phase I well under way, so I downgraded the QSN to a minor classification. Within 23 days, Phase I was humming along nicely and records of objective evidence were produced and attached to the QSN. I cleared the nonconformance without a revisit. A few weeks later the company had its ISO certificate beautifully framed and on the lobby wall. Three months later, Phase II was all set to go and ran in harmony with Phase I for a week or so. Finally, Phase I controls (manual cards, temporary person, outsourced calibrations, and so on) were eliminated, and Phase II became standard practice.

At the company's first routine maintenance visit by the registrar, (six months after award of the certificate), Phase II controls were evaluated and found to be operating perfectly. Another happy ending.

14 Administrative Processes: Going Beyond Manufacturing

If your quality system is in the formative stages and you are learning all you can about how your company works, this chapter will be especially helpful to you. Please read and apply it before sitting down to crank out your quality procedures. If your system is more mature, and your internal auditors are bored playing police officer and want to really improve the operation, this chapter is for them as well.

It would be cruel to bore you by telling you again how many assessments I've done in the past 10 years. Suffice to say, I'll bet coffee that it's more than you and your whole internal audit team have done together. In less than 20 percent of those audits did I command expert knowledge of the operational processes (of course, at least one other team member did). This limitation is true of virtually every registrar's assessors. Are you surprised? Be reminded that the standard assigns only one element directly to process control. Even if you include the few other elements that indirectly relate (inspection and testing, control of nonconforming product, quality planning, statistical techniques, and so forth), there are still more than a dozen elements that are generic to the quality system at large and are not process specific. Still, it is only the operational aspects of the business that are thought of as processes. All of the remaining activities, the "administrative stuff" to the manufacturing employees, are just something else. But consider this: Aren't all of these administrative activities also processes? By my definition they are.

A *process* is a value-adding activity that takes input (information just as legitimately as raw materials), acts on it, and converts it through the consumption of resources (time, money, people, energy, and so forth), then produces an output having greater value than that of the input and consumed resources combined.

As an assessor, I tend to evaluate these administrative functions (design control, management reviews, quality planning, contract reviews, document and data control, internal quality audits, training, and purchasing) as value-adding processes, not too unlike the way I evaluate traditional manufacturing processes. Actually (in personal life), I may go overboard a bit; my wife and kids assert that I sometimes do. Banking transactions, restaurant service, food store checkout, airport check-in and baggage processing, hotel reservations and check-ins, store merchandise returns, and my all-time favorite—vehicle registration—are all critiqued as processes having inputs, actions, and outputs. My motive is to identify inefficiencies and improve effectiveness. Can you imagine how much better these everyday events could be and how they could be improved if responsible parties would assess them as processes?

Most people don't stop to think of them as processes. Three times a year our company holds a tri-annual training event. The staff and assessors from around the country meet at an off-site conference center for two or three days of training. What a great way to run a registrar! I was in a breakout session, and our group was tackling continuous improvement issues. When I called the folks to order, I suggested that we define and adopt a process for how we should accomplish our mission. Three people looked at me as if I had two heads. A fourth team member just looked on in amazement. The idea of treating a team problem-solving, question-and-answer, and business analysis session as a process was foreign to all but a minority. Our team proved to be only marginally effective. So it's no wonder that a task like contract review is not readily seen as an administrative process.

Assessor Questions

When I conduct closing meetings at the end of unusually long assessments, I try to temper some of the negative findings with something like, "Well folks, I've looked at hundreds of pieces of objective evidence, made hundreds of observations, talked to dozens of people, and asked hundreds of questions. So the few items of concern must be viewed in light of the many, many positive aspects demonstrated throughout your quality system." More than once I've been tempted to have my guide/escort click a counter every time I ask a question. Ten or 20 an hour would not be an exaggeration.

Asking honest, sincere questions is one of the best tools an assessor has for collecting information. Asking the right questions permits the auditor to get up to speed quickly and to accurately evaluate a process in minimal time. While registrars do not to seek to critique people, it is through people that the quality system is implemented and sustained. Assessors must ask questions of people. The second best tool is for assessors themselves to ask questions about observations they make and about the objective evidence they see, hear, and touch.

I do not know your company or your people. I don't know your quality system, and I certainly don't know your administrative processes. So how can I best equip you to analyze the nonmanufacturing side of your business? Suppose I give you a list of the questions I might ask (or think) if I were personally assessing your administrative processes on-site.

You may have heard that when an assessor gets an answer to a question, he or she will ask another question directed at the previous response, and then another question following every response. Kind of like the child that keeps asking Mommy, "But why? But why? But why? . . ." When the assessor has exhausted all of the questions and the auditee runs out of responses, then the assessor has learned all he or she needs to know about the subject at hand. Well, I have lots of questions. You may wish to use them when conducting your own internal audits or subcontractor audits.

For presentation's sake, I'll order the questions much differently than if they were actually being asked. Think of four categories: who, input, action, and output.

What are the *whos* associated with this administrative process?

- Who owns the process?
- Who is responsible for the process?
- Who champions the process?
- Who facilitates the process?

- Who coordinates cross-functional involvement?
- Who actually executes the process?
- Who gets credit when the process works well?
- Who gets criticism when the process doesn't work well?
- Who is positively influenced by the process?
- Who is expensed by the process?
- Who is authorized to make changes to the process?
- Who is responsible for providing input to the process?
- Who is responsible for producing each process output?
- Who actually produces each process output?
- Who wants the process output?
- Who needs and should have the process output?
- Who is actually using the process output?
- Whose turf is jeopardized by process change?

What are the *inputs* associated with this administrative process?

- What are all the individual process inputs?
- Can each input be classified as critical, major, or minor?
- What process inputs are controllable?
- What process inputs are not controllable?
- What uncontrollable inputs can be reduced or eliminated?
- What controllable inputs can be improved now/short term
- What controllable inputs can be improved later/medium to long term?
- What will it take to improve the process inputs?

    —Replace source?

    —Correct source?

    —Secure alternate/additional source?

—Establish/clarify input requirements?

—Enhance input requirements?

—Assess inputs before using in process?

—Rework less-than-perfect inputs

—Scrap less-than-perfect inputs?

—Make a temporary process adjustment to use imperfect inputs?

What are the *actions* associated with this administrative process?

- What in the quality system prescribes/requires the process activity?
- What makes the process go—what's the engine/driver?
- What are the macro activities in the process?
- What are all the micro activities in the process?
- Who performs each micro activity?
- What is the processing time for each micro activity?
- Which micro activities should/must be done sequentially?
- Can each macro activity be classified by importance/criticality?
- Can each micro activity be classified by importance/criticality?
- What is the consequence(s) if the process activity is flawed?
- What is the consequence(s) if the process activity breaks down?
- Do fail-safes or controls adequately contain failure modes?
- Are failure consequences commensurate with activity controls?
- Which micro activities should/may be done in parallel/concurrently?
- Are any micro activities redundant?

- Can any micro activities be combined?

- Can any micro activities be eliminated?

- Should some micro activities be done outside of the process?

- Should some external activities be brought into the process?

- Are adequate resources assigned to each micro activity?

- Are personnel adequately trained and equipped for the task?

- Are metrics associated with each macro activity?

- Are targets and goals associated with each macro activity?

- Is each macro activity critiqued against expectations?

- Is overall process performance evaluated against a quality objective?

- Is the suitability and effectiveness of the overall process activity subject to formal management review?

- When and how is the need for process activity change identified?

- How are process activity changes initiated and implemented?

- How are new process micros identified, initiated, and implemented?

- What are the short-term process improvement actions?

- What are the long-term process improvement actions?

What are the *outputs* associated with this administrative process?

- What is the overall objective of the process?

- What is the output from each micro activity?

- What is the value of the output from each micro activity?

- What is the output from each macro activity?

- What is the value of the output from each macro activity?

- What are the metrics associated with each micro and macro output?

- What is the quality objective associated with the process output?

- What is the performance expectation with the process output?

- What are the target(s) and goal(s) of the process output?

- Are any process outputs redundant?

- Are any of the process outputs lacking?

- Can any of the outputs be combined?

- Can any of the outputs be eliminated?

- Can the output frequency be reduced?

- Will outputs later be combined or consolidated?

- Should some other process produce part of the output?

- Does each output justify its input and action cost?

- What is the relative importance of each output?

- Are key process outputs subject to formal management review?

In addition to posing and answering questions, another very effective tool in understanding and improving administrative processes is the use of flowcharting. Flowcharts can be further used to prepare administrative procedures. You may even choose to include the flowcharts within the procedures themselves as an overview aid. Flowcharts effectively expose redundancies, disconnects, and wasted loops and identify unresolved conditions. As you tackle each administrative process, consider preparing a macro flowchart, complemented with one or more very detailed micro flowchart. Operators of the process may be the most qualified to provide the nitty-gritty details. Someone knowledgeable of flowcharting principles can hone them into an invaluable aid.

If you, your audit team, and your administrative process operators will take the time and effort to ask these questions, everyone will learn more about the "soft" side of your business than you ever imagined. If what is learned is taken to heart and applied, substantial progress can be made. Once a process is defined and understood, it can be brought under control. Once it is controlled, you can begin to improve the process—be it operational or administrative.

15 Connect the Dots with Interface Agreements

Alhough it has been more that a dozen years since my children were little enough to order from the kiddy menu, I can still recall vividly how they loved to play connect-the-dots on the back of the restaurant placemats. The placemats were great at occupying our kids while our dinners were getting cold in the kitchen. Even today, I am tempted to work these little teasers when they occasionally appear in the children's section of our church bulletin. Perhaps it is really the rush of nostalgia that I succumb to. Now honestly, tell me you haven't. By simply drawing a visible line between two adjacent dots, and then on to the next dot and the next, you can create a picture of some magical scene or character. What was once just a bunch of independent, fragmented, and disassociated black specks has now become a meaningful representation. You can now begin to appreciate the relationship between the dots, which was not previously discerned. It becomes clear just how each dot plays an important role in the overall scheme. Each dot gains strength from its neighbor so that the sum is greater than its component parts. So it is, too, with some quality systems. Call them what you will—connect-the-dots, interface agreements, MOUs, or corporate interlinks—there exists a very real need in companies to document and clearly represent the relationships between various business entities and their activities. This is especially true in some quality systems and particularly in larger, multisite environments where corporate or divisional groups service the plants.

There are many such services that typically include sales, purchasing, marketing, R&D, servicing, design, calibration, engineering, customer complaint processing, contract review, warehousing, internal audits, control of standards and codes, training, information management services, personnel, laboratory support, and health and safety. For many companies it makes perfect sense to centralize these activities rather than to reinvent the wheel at each production facility. Redundant resources and efforts can sometimes be eliminated while maximizing consistency and efficiency. When centralization is the approach, the activities of the quality system must still be defined, documented, controlled, and compliant with all applicable requirements in the standard.

Keep in mind one very important stipulation: The registered site (the one to whom the ISO 9000/QS-9000 certificate is to be issued) is responsible for satisfying all applicable elements of the standard, whether they are performed at/by the site or at/by a corporate/divisional group. Let's go back to my opening example. Even if some of the dots are at the plant, some are at a corporate/division location, and maybe even some are at a third company site, all of the dots must be appropriately connected if the picture of the quality system is to look like Beauty and not the Beast.

Often I've heard the exclamation, "Oh, our purchasing division takes care of all purchases, and it chooses all the vendors too. It even handles vendor quality problems. This plant just sends in receiving inspection data and purchasing does the rest at headquarters." Then I pose a number of questions like the following:

- How are the subcontractors selected and approved?

- Does headquarters or purchasing ever audit the vendors?

- Do the plants get reports of vendor performance?

- Is the receiving inspection data actually used?

- How is corrective action applied to the subcontractors?

- Are purchasing requirements reviewed and approved before release?

- How does the plant know if purchasing and headquarters are following the pertinent requirements in the ISO 9002 standard?

- Is purchasing adhering to document and data control and to training requirements?

Finally, I ask the site quality system management representative, "Did you know that even though you don't do purchasing at this plant, you are still responsible for the activity—even though headquarters is performing the service on your behalf, you need to ensure that it is properly and consistently done?"

"Huh! No, I didn't," is the common response. "Besides, how would we know what they do over at the main office?"

"Precisely my point," I say with just a bit of pleasure because now the rep is starting to ask questions. "Have you ever thought about writing down the relationship between you (the plant) and headquarters? Have you considered a document that describes exactly what purchasing service they provide (who, what, where, when, why, how) for you? It might even describe how the effort will be controlled and evaluated. It may even be reviewed and accepted by both parties so that everyone is clear about expectations."

When preparing interface agreement documents, remember that they must accurately and completely represent the service being provided as well as the relationship between the provider and the recipient. However, their content, format, level of detail, and formality can vary widely depending on several factors.

- Structure of the organization
- Company culture
- The criticality of the service being performed
- Task difficulty
- Necessary level of control
- Ease and convenience of oversight activity
- Frequency of the service
- Impact on or involvement by a third party
- Existing company procedures and work instructions
- Specific requirements of your registrar, company, and industry

To further understand the value and purpose of an interface agreement, consider the relationship between your company and one of its key suppliers or subcontractors. Where once a verbal description of the purchased item followed by a sincere handshake was more than adequate, today's business environment requires requests for quotations, signed purchase orders, technical specifications, and terms and conditions. Very little can be assumed or taken for granted. Both parties must mutually agree upon what is expected by the purchaser, and what will be provided by the supplier, and under what conditions, and in compliance to what quality requirements, and at what price, and when, and . . . Resultant descriptive documents are intended to eliminate confusion, ambiguity, and poor customer/supplier relationships (broken promises, hurt feelings, and lonely hearts).

Now apply this example to internal relationships in your company. Virtually every company has its own internal customers and internal suppliers, and what takes place between them is no less important than that between the company and external providers. The quality system registrar will expect to find supporting documentation for these internal relationships—documentation not unlike that of external relationships, say, between you and your customers (contracts and orders) or between you and your subcontractors (purchase orders and specifications). The registrar must be confident that all applicable elements and requirements of the standard are adequately addressed and satisfied, regardless of who is performing the activity—be it the local site/plant, a sister plant, a corporate group, or headquarters. It is crucial to remember that the location/entity being registered and certificated is ultimately responsible for the activity—its accuracy, completeness, compliance, control, and record keeping. It has to make sure that all the dots are connected.

I have seen interface agreements used very successfully in the largest of manufacturing companies, as well as in the smallest service companies: from computers to consultants, from distributors to R&D centers, from textile finishers to software houses.

Now here's the icing on the cake (and this is from actual personal experience with a registrar): If the relationship and activity is adequately defined and documented, and if there is adequate control exercised over the service provider by the recipient, and if there is sufficient objective evidence (for example, quality records) that the entire process is meeting all essential requirements (standard, industry, company, regulatory, statutory), then it is quite likely that the registrar will not actually visit the location of the service provider (at least not every time), just as it would not visit your subcontractors. The registrar still reserves the right to visit the service provider if the interface agreement is inadequate or if there is evidence (or strong suspicion) that all is not well.

Following is a list of components (not in any particular order of importance) that you should at least consider including or stipulating in your interface agreements.

- Required quality records that are to be produced and maintained
- Reporting criteria, mechanisms, and distribution
- Reference supporting procedures (corporate, site, and so on)
- Signatures of participating group representatives
- Dates and duration of agreement
- Detailed descriptions of services to be provided
- Special terms or conditions
- References to applicable elements in the standard
- Notification of changes to pertinent reference documents
- Required approval of changes to pertinent reference documents
- Key representatives and contacts for the agreement
- Internal audit activities by the service provider
- Participation in the audits by the service recipient
- Corporate audit activities over the service and provider
- Distribution of audit results to service recipient
- Control and oversight scheme
- Service performance criteria
- Specific quality system requirements
- Suspension and termination conditions of the agreement
- Periodic review and reapproval of the agreement
- All the who, what, when, why, where, and how stuff

If you will take the time and effort to assemble all of the dots and then visibly connect them in the proper order, you may see something magical appear before your very eyes: a well-documented, disciplined, and effective quality system.

16 Identifying the Need for Statistical Techniques

The ISO 9000 series standards and QS-9000 requirements include a requirement relating to the identification, implementation, and control of statistical techniques. While QS-9000 is quite specific and somewhat prescriptive, ISO 9001 and ISO 9002 are much more vague about what is required of the supplier. The 1987 version of the ISO documents was even worse. Essentially, the 1987 version created a giant loophole through which the majority of companies tried unsuccessfully to leap—with a capital L.

The standards used a very wishy-washy phrase: "Where appropriate, the supplier shall . . ." This was all-too-often and incorrectly interpreted by the supplier as "If you want to . . ." or "If you would like to . . ." or "If it is convenient to . . .". Needless to say, registrars saw minimal application of genuine, meaningful statistical techniques. Instead, the supplier flatly declared that "they're not appropriate," though what they were really saying was, "We don't like them, don't understand them, and don't want to be bothered with them."

In the 1994 version of the ISO 9000 series standards, a significant clarification was asserted, and the "where appropriate" marshmallow stuff was removed—appropriately. Now the ISO 9000 documents specify two different, but associated, requirements for statistical techniques: identification of need and procedures, as follows.

4.20 Statistical techniques

4.20.1 Identification of need
The supplier shall identify the need for statistical techniques required for establishing, controlling, and verifying process capability and product characteristics.

4.20.2 Procedures
The supplier shall establish and maintain documented procedures to implement and control the application of the statistical techniques identified in 4.20.1.

Most registrars agree—with much relief—that the "shalls" carry considerably more weight. The words in the standards now closely align with the interpretations commonly held by registrars all along. Allow me to offer a personal elaboration on element 4.20.

ANYTIME YOU

Make inferences   Draw conclusions   Apply assumptions

ABOUT

Qualifications   Capability   Accuracies

Variability   Predictability

OF

Activities   Products   Processes

Equipment   Services

THEN

There is at least an opportunity to identify potential needs for statistical techniques and there is then the responsibility to control their application through documented procedures.

Now that I have set the potential scope of applicability for your statistical techniques initiative, and you are undoubtedly sufficiently overwhelmed with its enormity, let's see if there is some practical, manageable approach to bringing the beast to its knees. I just happen to know of one that has been effective for several companies—a modification of the do-it-yourselfer's home maintenance project.

Step 1: Assemble Your Toolbox and Supplies

Identify the various statistical techniques that are available to your company and your industry. They will include one or more of the following (not all are considered true, conventional statistical techniques in the purest sense):

- Design of experiments (DOE)
- Taguchi methods
- Process failure mode and effect analysis (Process FMEA)
- Product FMEA
- SPC
- Process capability indices
- Gage repeatability and reproducibility (R&R) studies
- Statistical sampling
- Probability calculations
- Run charts
- Precontrol charts
- Predictability calculations
- Pareto analyses and histograms
- Quality function deployment
- Reliability calculations
- Mean-time-between-failures calculations
- Mean-time-to-repair calculations
- Confidence limit calculations
- $\bar{X}$–R charting

Step 2: Review and Prioritize Your "Honey Do" List

This exercise will take some time and hard work, and should involve good cross-functional representation by several groups in the organization such as design, R&D, analytical engineering, process engineering, product engineering, manufacturing, quality engineering, reliability, customer service, and marketing. It is very important that the group take excellent notes and document all of its efforts (who, what, when, why, where, and how). Let's refer to this group as the needs assessment team. Its mission is to thoroughly evaluate every aspect of the business relating to product characteristics, process capabilities, and yes, even administrative activities, for every opportunity to employ any of the statistical techniques in your toolbox. You may be surprised at just how many there are. Be open minded. Take off the blinders and think far beyond traditional SPC. Once the team has identified all the candidates (perhaps 10, 20, or more), they need to be prioritized. Recognizing that no company has unlimited resources (time, talent, personnel, money, and production capacity), rank items on the list by using the following criteria (to name a few).

- Cost to apply the technique
- Anticipated return or benefit
- Difficulty in applying the technique
- Time to implement the technique
- Impact on operations
- Industry and customer demands
- Cost of not using the technique
- History of problems or failures
- Challenges from competition
- Quality policy and objectives
- Organizational goals

Upon completion of Step 2, and with the finalization of a prioritized list, you have nearly satisfied the identification of need discussed in paragraph 4.20.1 of the standard.

Step 3: Match the Tools to the Tasks

Start at the top of the list and work down until you have exhausted the resources devoted to your company's statistical techniques initiative. For each need on the list, scrounge through your toolbox and choose the best, most appropriate statistical technique for the job. Just because a pair of pliers will remove a hex nut does not mean they will do a better job than a box wrench. My dad used to say, "A good mechanic is known by his tools, and only a jerk would use pliers on a hex nut." You can't believe how many times I've seen companies use the wrong statistical technique—like process control charts on an out-of-control process. This creates more problems than it solves because of unnecessary process adjustments (tampering). Another frequent mistake is calculating and then applying process capability indices (C_{pk}) when the data are not normalized.

Step 4: Prepare Your Workbench

Secure the necessary instructional aids and reference books. Learn the correct methods for applying the selected statistical technique(s). Develop an implementation plan. Get management support! Convince the "turf lords" that the rewards will surpass the pain. Do your homework. Plan, prepare . . . plan, prepare . . . plan, prepare . . . then plan and prepare again! I can't support this claim statistically, but I'll bet most companies fail three times before they successfully implement a genuine SPC program. Ill-conceived, poorly planned, and half-heartedly implemented initiatives only serve to turn off personnel and make the next uphill climb a little steeper. Clear off the workbench and get rid of the debris, junk, and greasy tools from the last job. Learn from—but put aside—past mistakes.

Step 5: Latch onto Your Teenage Helper

My son is usually a good helper. But he becomes a great helper when he is motivated, stays focused, and can see some reward or benefit in doing the job well. I contribute the training, tools, spare parts, and coaching. By keeping his stomach full and hiding the cordless telephone, I further improve chances of success. Identify those individuals in your organization that have the know-how, time, resources, initiative, and vested interest in seeing successful implementation. I've learned that sometimes it is best to assign responsibility for implementation to the party that has the most at stake or the most to lose if the process fails.

Once upon a time there was a wise old quality assurance knight that struggled for months trying to install an in-process inspection program in Widget Land. The kingdom had always relied on a toll-gate mentality and was therefore burdened with a huge final inspection army. The knight had the right idea, but the lord of manufacturing wanted no part of it. It was very convenient for manufacturing to let all of the widgets (conforming or not) escape into Final Land, where it became "their problem." The stuff poured into Final Land like a tidal wave at the end of each month. Alas, the plant king awoke, but only after the knight refused to ship anything that was not a perfectly beautiful widget for a couple of months. The king painfully realized that somebody's crown jewels and the very throne were in jeopardy. The battle turned overnight. The floor inspection infantry was reassigned from quality to manufacturing. Manufacturing was charged back for any nonconforming widgets found in Final Land. This is the best part: Inspection costs and overhead were also charged back to manufacturing for nonconforming widgets that crept into Final Land. Add that to the cost of repair/rework, and it was easy to see why the manufacturing lord became determined to effectively utilize his new-found inspection troops. Within six months, the kingdom realized a 23 percent reduction in scrap and rework costs, and everyone marveled at the 12 percent savings in appraisal costs that resulted from downsizing the final inspection army. Returned goods and customer complaints were rarely heard of again throughout the kingdom. It's amazing how a new owner can dress up an old castle. The end.

Step 6: Take Notes and Pictures So You Can Do It Again

At this step you are ready to close in on paragraph 4.20.2 of the standard: statistical technique procedures. This can be a fairly routine, mechanical effort provided you have properly completed steps

1 through 5. The notes and documentation from Step 2 will come in very handy. System-level procedures, departmental procedures, and/or work instructions must be defined and documented. They must address the who, what, when, why, where and how of the statistical techniques, the areas of application, and methods of implementation to ensure proper usage and control. Gage R&R studies are a common example of incorrectly applied statistical techniques. Companies will fail to use more than one inspector, or they unknowingly bias the results by letting the inspector(s) inadvertently know the actual values of the measured parts. The outcomes of such R&R studies are meaningless. The documentation should also describe the needs assessment process so that it can be effectively repeated as needed, should business conditions change (as they will) in the future.

When my son and I built my workshop, did all the plumbing and electrical work on the new in-ground swimming pool, and converted the screened porch into a family room (over the course of 15 years and three different homes), we were careful to have formal, detailed drawings and schematics (which we prepared ourselves and had approved by the city). We also took 100 or so color photographs throughout various stages of each project. These are great crowd-pleasers that I show to all new guests at our present home. My wife is annoyed by this practice, and my son is both embarrassed and proud. These materials, besides being great reminders of our early family years, has proven to be very useful to my friends when they have some questions about how to perform their own projects and want to avoid making mistakes. The quality system documentation on your statistical techniques will be just as useful to your workforce.

Recall that I am somewhat critical of the 1987 standard's use of the phrase "Where appropriate, the supplier shall . . . " as it related to statistical techniques. However, a reasonable registrar must be open to the possibility that in a few cases, for some companies, statistical techniques may not be appropriate. (I can hardly believe I said that.) When performing assessments, this possibility must be taken into account, and the registrar must be open to the supplier's (your company's) explanation. If the explanation is just smoke and mirrors or just a replay of the old, worn-out song, "We don't like them, don't understand them, and don't want to be bothered with them," then the supplier has not done its homework, and is not compliant with the requirements of paragraph 4.20.1 in the standard.

On the other hand, if I hear the supplier sing a snappy little tune with the lyrics, "Oh, statistical techniques are truly not appropriate, truly not appropriate, truly not appropriate . . . We have reached this conclusion only after a thorough needs analysis of all our business activities, products, and processes. We investigated various statistical tools and determined that none are applicable for several justifiable reasons which are . . . We have some detailed documentation that substantiates and validates our position on this—would you like to see it?", then I can hardly resist tapping my feet and snapping my fingers.

Now do you see why good notes and documentation are so important? If the documentation supports the supplier's assertions, then I am comfortable and assured that paragraph 4.20.1 has been adequately met. Furthermore, if no statistical techniques have been identified (bona fide), then the requirements in paragraph 4.20.2, Procedures, are irrelevant.

CHAPTER 17

Supporting Documentation for Design Control

Design control is an especially interesting element of the standard to me. While I may not be an expert on a company's specific products (for example, computers, medical devices, automobile tires, diesel engines, or electronic scales), nor on its technical attributes and performance characteristics, I am quite comfortable with the organization, structure, discipline, and documentation necessary for a good design control program. Irrespective of the particular product being designed, there are basic, essential qualities associated with an effective design control system. I want to emphasize documentation, procedures, work instructions, forms, and especially quality records. This emphasis is the thrust of this chapter. For those of you who enjoy taking vacations by automobile, this one's for you.

Pre-assessments performed by a registrar are customarily formatted to evaluate all elements of the quality system, like a dress rehearsal of the formal registration audit. But sometimes a customized, focused pre-assessment is in order. One of my larger clients has elected to do just that for the R&D and design centers for several of its business units. On three occasions the company has asked me to concentrate on the design control activities at each site. In each of these three cases, I was fortunate to have one-and-a-half days at each center. Compare that to the one to two hours typically devoted to a given element during a traditional assessment. This is an assessor's dream come true. Finally, I did not have to be content with a limited sampling process. There was ample time to thoroughly evaluate the entire process in glorious detail! Virtually every aspect of the requirements of the standard could be examined. I wanted to use my time and knowledge in a way that would maximize the value derived for the client, to best prepare it for its upcoming registration assessment, and to acquaint it with every possible question that might be raised later. My biggest challenge was to ready it to be able to produce necessary quality records to demonstrate that it satisfies specified requirements (internal, customer, industry, standard) and the effective operation of its design control program.

I travel a lot, mostly in the southeast, but occasionally as far north as Nova Scotia or as far west as Iowa. A prerequisite is a good road atlas and a detailed map from the client. My travel agent and the airlines reliably get me into the closest airport, but from there on I'm on my own—the rental car does not know the way. Unfortunately, clients sometimes neglect to put compass orientations and/or mileage on the maps. Sometimes they call out quaint indicators like "the yellow house with white rockers on the front porch." To be sure I'm on course and on time, I must identify specific distances, mile markers, landmarks, highway intersections, and the like. If these indicators appear at the right time and in proper order, chances are the intended destination is just around the next corner. It may be 50 miles from nowhere and in the middle of a cornfield in Illinois, or dangerously close to no-man's-land in Yonkers, New York. I love it when a plan comes together.

A journey (assessment) through a design control program is not much different. There must be a plan—a road map, so to speak (procedures and work instructions). There must be directions, timetables, and key milestones (design/development plan). You must be able to determine that you have successfully passed one intersection (design phase) before proceeding to the next. You must be able to demonstrate that you are on time and on course. For the design control journey, that means objective evidence (quality records), required approvals, predetermined handoffs, essential input and output for each phase, and performance criteria. Without these disciplines, problems—if not failure—are likely to occur. Imagine a relay race where the fourth runner crosses the finish line well ahead of the closest competitor. What must go through his mind when he realizes that runner number three forgot to pass the proverbial baton? The thrill of victory. The agony of defeat!

I have seen more than one design validation on an expensive prototype end in failure simply because crucial erroneous assumptions were not investigated in early design reviews and were not caught at subsequent design verifications. There have been calamities associated with failure to obtain mandatory customer approvals before making process or product changes. Lawsuits occur when products do not perform because the design team concerned itself with only critical and major characteristics during the design FMEA. The results to these design control problems equate to being lost in the middle of an Iowa cornfield at 10:49 P.M. with a gas tank down to all but fumes and the beginnings of a heavy snowfall.

Join me and my family on our vacation trip along the scenic design control highway. Be mindful that there will be some dangerous construction areas (critical and major characteristics requiring design FMEAs), blind hairpin curves in the mountains (unplanned customer design changes), several toll booths (design reviews), at least one highway patrol roadblock (design verifications), and, of course, the final border crossing into Canada (design validation). Despite predictions of some nasty weather in the northeast (missed design and planning schedules), the remote possibility of a freak mechanical problem (faulty design assumptions) even though the car was just checked out and serviced by the rental agency, and an overly aggressive hitchhiker (a design manager who despises formal documentation), the trip promises to be filled with great views, fine restaurants, comfortable hotels, and overnight reunions with close friends, and will finish with a well-deserved week at a wonderful lakeside resort.

Before I even leave the driveway, though, I need to study the road atlas and maps. There must be a clearly defined and documented description of the design control program, starting in the quality assurance manual and threading its way through the system-level procedure(s) and low-level work instructions. During the audit, these documents must provide an explanation of the workings of the program (who, what, when, why, where, and how). These maps must be current, otherwise they may not reflect current practices (interstate extensions, new connecting roads, and so on). These maps also need to specify reference documents, forms, and essential quality records (rest areas, scenic overviews, toll points, construction areas, and eating and refueling stops). I will periodically refer to these guides during the assessment just as I use maps on our trip. This constant referral to the procedures is like listening to the radio along the way. The radio is a great companion, helps keep me alert, and notifies me of inclement weather conditions and any congested traffic or accidents ahead.

Our imaginary vacation trip actually started with an invitation (request for quotation) from relatives in Canada. We also received some complimentary hotel accommodations (thanks to all of my guest points accrued over the years) that have some restrictions (design input criteria), and a seven-day car rental discount coupon that also carries its own special restrictions (customer drawings and specs). Friends along the way have insisted that we stop by for lunch or to spend the night—but there are strings attached to these obligatory visits. Their kids are brats, they want to borrow money, and they show us boring home videos (external standards and codes). But we really do love our friends (paying customers), and we are glad to stop. Well before our departure, I made confirmed reservations and notified friends and family along the way, accepting their invitations and communicating the details (proposals and project scope).

On the Monday before the departure weekend, I sense that it is time to enter my planning mode. My family accuses me of being a fanatic. I prepare detailed checklists of items to bring and things to do (design/project plan), assign my daughter the job of farming out the pets (subcontractor management), have my son service the vehicle, ask my wife to cancel the newspaper, and reluctantly beg my nosy neighbor to pick up the mail and keep an eye on the house (organizational and technical interfaces). I know exactly when to do certain things each day during the week (milestones and timelines). Each person is selected for the tasks to which they are best suited (qualified personnel assignments).

On Friday night I even do a trial-run prepack of the trunk to be confident everything will fit the next morning (design feasibility study/concept review). My wife and I huddle around the dining room table and analyze the chosen route for the excursion. After a thorough evaluation of this crucial component of our trip, we mutually agree on its acceptability and shake hands (design software approval). The route is carefully chosen based on several prerequisites (design input criteria): We must stop at the homes of every friend and relative along the way. The wife, kids, and I have picked four favorite tourist attractions. Primary highways and interstates are used whenever possible.

Finally, Saturday morning arrives and the plan is coming together nicely. Each family member is assigned specific tasks and has a role throughout the entire trip (personnel and resource allocations). Number-one teenage son is responsible for checking the oil, tire pressure, and cleaning the windshield at each stop. Number-two adolescent daughter will replenish the snack bag at every service plaza and will compile an assortment of postcards, souvenirs, and tourist attraction brochures. My wife will be chief navigator, official backseat driver, and referee for sibling fights. I make sure we have essential supplies (design tools, FMEAs, DOEs) like road flares, a first-aid kit, the toolbox, plenty of tape cassettes for my personal listening pleasure, and a can of instant fix-a-flat so my son doesn't have to change a tire in the rain. With my wife supervising, I load the trunk. It's necessary to break out the car-top turtle shell because of all the unplanned extras and last-minute junk that we can't live without (internal design changes).

Don't be misled. I'm not all work and no play. Everyone in the family has been given the okay to spend money, and we all know what we are supposed to buy (internal work requests and authorizations). There will be vacation wardrobes, swimsuits, fishing gear, gifts for the friends and in-laws, and games to play in the car. Everyone gets an appropriate amount of money for discretionary spending, too.

Just before we get into the car, we convene in the family room for the first official powwow. We run through the checklists, making sure we didn't forget any task assignment. We check and recheck the things-to-bring list. My wife makes sure we have plenty of cash, phone numbers and addresses, and hotel confirmation numbers (first design review). We rehearse the route, order of visits, attractions, schedules, everything. The whole family agrees once again—and the official plan is cast. We are all committed to having a good time. We pause for a moment of much-needed prayer, give each other a hug (project plan review and approval), then dash to the car. I love a good planning exercise.

At the end of our first day, we discuss over dinner how the trip is going. This is our first real critique of the plan as it is unfolding (design review). We make a few refinements like alternating seating assignments and agree to take longer rest breaks, more frequent potty stops, and less candy and more crunchy snacks. Everyone is satisfied with the improvements (design modifications and approvals). I'm careful to remind the gang that the extra, longer stops will mean a schedule adjustment. Either we add an hour to each day's travel time, eliminate a planned attraction, or forget seeing Aunt Betty. We compromise by adding 30 minutes and dropping Aunt Betty (project plan review, update, and approval). After the second full day on the road, we evaluate nearly everything. The car is performing well, preselected restaurants and hotels meet expectations, we are covering enough miles, no one is overly tired, the chosen routes have been construction-free highways and interstates, and calls back home find the house and pets doing fine. This favorable critique (design verification) proves our vacation plan is coming off without a hitch. The kids inform us that they have been keeping a diary on how well each attraction, meal, hotel pool, game room, tourist trap, and family visit rates on a scale of 100. They check each item against 10 or so attributes. Their official findings (test results and approvals) will decide what makes the cut and will be tried again next year.

When we pass into Canada, the customs officers question our purpose for crossing the border, examine the contents of our turtle shell carrier, confirm and stamp our passports, then signal us to proceed (more test results and approvals).

At the end of the eighth day we finally arrive at our destination: a ski resort on beautiful Lake Dee-sign-a-control-ee. Three months earlier I had purchased a family vacation package that included a master suite, meals, water-ski tickets, local transportation, horseback riding, greens fees, a moonlight cruise, and unlimited tokens at the video game center. Everything was described in a signed contract between me and the resort (customer contract/order). We received color brochures showing the rooms, the clubhouse, a panoramic view of the lake, the Friday night seafood buffet spread, and local antique shops. This sold my wife on the resort (prototype approval).

Upon check-in, the resort director begs my forgiveness as he explains that horses are sick, the middle eight golf holes are under water, and the sheriff has confiscated the video games because of illegal gambling by the local juvenile delinquents. He assures us that we would be amply compensated by a price reduction, an extra two days of lodging, an upgraded room, and so on. I accept the conditions, and the director has the terms of our contract revised and has me initial it to show my agreement (approval of customer waivers and exceptions). But first, the family scopes out the facilities, rooms, rowboats, and so on to be sure everything else is as we expect (another design verification and test).

On the last day of a fantastic stay at Lake Dee-sign-a-control-ee, the activities director asks the family to complete a very special customer satisfaction survey. The hotel needs no less than 25 such surveys with outstanding scores to qualify for a four-star rating from the area Chamber of Commerce. We agree and give the scores they are hoping for (production part approval).

We returned home the following Saturday evening to a good night's sleep in our own beds and an excellent home-cooked breakfast on Sunday morning. A week later we get back 100 or so pictures that we took at every stop and with every friend and family member. After dinner on Wednesday, the clan gathers in the living room to reflect on our vacation. There were six specific objectives for our trip (design output criteria). In order of importance they were (1) arrive home safe and sound, (2) strengthen family bonds, (3) rekindle friendships, (4) have lots of fun, (5) rest and relax, and (6) experience new places and things. We look at every photo, recall special moments, and laugh, and the kids count their leftover money. We talk about how well we met our vacation goals and agree that all the up-front planning really paid off (design validation). The maps, checklists, timetables, attraction brochures, photographs, addresses and phone numbers, hotel menus, bills and receipts, and kid's diary (quality records) are tucked away in a safe spot so we can refer to them when planning for next year's trip (new/repeat orders).

A real design control initiative is not much different. As an assessor, I expect to see a number of supporting documents on file. These will give objective evidence of effective implementation of quality system procedures, compliance to requirements, and effective operation of the design control program. The following is a checklist of those documents that the registrar will expect to see (if applicable). Share the list with your design personnel and internal auditors. Make sure everything is available. And oh, by the way, have a great trip!

Supporting Documentation for Design Control Activity

- Quality assurance manual _____

- Quality system procedures _____

- Quality work instructions _____

- Customer requests for quotation _____

- Referenced standards and codes _____

- Design product specifications _____
- Test performance specifications _____
- Customer prints and specifications _____
- Internal work requests and authorizations _____
- Definition of project and scope _____
- Project plan and timetable _____
- Personnel and resource allocations _____
- Organizational and technical interfaces _____
- Design software approvals _____
- Design input requirements _____
- Design project phase approvals _____
- Design review records and minutes _____
- Design output requirements _____
- Design verification approvals _____
- Test results and approvals _____
- Design validation approvals _____
- Project plan reviews and updates _____
- Design change requests and approvals _____
- Design tool records (DOE, FMEA, and so on) _____
- Design tool results and approvals _____
- Subcontractor approval and oversight _____
- Customer approvals of designs and changes _____
- Customer approvals of tests _____
- Prototype approvals _____
- Production part approvals _____
- Customer approvals of waivers and exceptions _____
- Acceptance criteria for all design phases _____
- Prescribed internal forms and quality records _____

18 Management Review Summary: A Real-Life Example

After reading chapter 6 on management reviews, you may be thinking that some of what I've presented is pie-in-the-sky and that real companies don't even come close to management reviews such as I've been discussing. While I admit they are not the norm, many companies I've assessed have gone the extra mile. Some have great reviews and continue to improve them. One of the most rewarding times I have as a lead assessor is when I see the light come on in managers' eyes, knowing they have truly grasped the concept of management review and have effectively applied it to their quality system.

One such company is Saft America–Industrial Battery Division in Valdosta, Georgia. Saft's director of quality, Gary McMillan, has graciously permitted me to reproduce the actual summary report from one of the company's management reviews. As you will see, Saft takes quality very seriously. Saft's employees know that the quality system is a key to their success. They know that management reviews are essential to the health and progress of their system. Saft employees conduct their reviews three or four times a year. It is a high-level review by staff and senior management at the facility. As I recall, it is a day-long, off-site event. Saft's review documentation is among the best I've seen.

Pay particular attention to how Saft cites not only its strengths, but also its weaknesses. Saft's employees know it is *their* system being reviewed, and they want to improve it. See how they assign action items and responsible people. Look at the deliberate, systematic, step-by-step approach taken to accomplishing all agenda items. Marvel at how they have set clear, measurable objectives and targets and then monitor performance against them. Note that they actually learn from reviewing all the input data—that they come away with tangible results and conclusions.

Is Saft IBD's quality system management review perfect? Is the documentation flawless? McMillan will be the first to say no. But I believe it is head and shoulders above most. If you're thinking of exactly copying Saft's, please don't. Your company, its organization and culture, and its quality system are special and unique. Take the time and effort to develop your own style and format for your management review meeting and minutes.

There are a few suggestions that may make your reviews that much better.

- Be sure to document the process in a procedure. Describe the who, what, when, where, why, and how of the management review.

- Identify attendees, frequencies, core agenda items, content of the summary report, routine input data, and responsibilities.

- In the output documentation, assign follow-up and/or required completion dates for every action item.

- Record the attendees and note those missing from the required/core attendance list.

- Mention the time, date, and place of any specially planned interim follow-up meeting.

- Have an authorized individual (probably the ISO management rep) sign and date the summary report or minutes to show they have been reviewed and approved and to make them official.

- Throughout the report/minutes, identify specific supporting data (name, number, date, printout identification, and so on).

- Declare the output report/minutes as official quality records.

Management Review Summary
November 3, 1995

  I. Internal audits

  II. Customer complaints

  III. Returns

  IV. Nonconformances and corrective action

  V. Preventive action

  VI. Progress toward each quality objective

  VII. Progress toward quality policy

 VIII. Organization structure, adequacy of staffing and resources

  IX. Performance, suitability, and effectiveness of quality system

  X. Management review summary

I. Internal Audits

Strengths

1. Vigorous audit program continues to identify system weaknesses.

2. New/additional auditors trained and functioning.

Weaknesses

1. Timeliness of response to discrepancies requires improvement.
Action: Review and summarize response time to quantify the issue. Report results for decision. Twenty-three open deficiencies. Expedite closure of open statements.
Responsible: L. Kolb

2. Timeliness of reviewing and closing out audits requires improvement.
Action: Review and summarize the time to quantify the issue. Develop regular two-week reporting cycle.
Responsible: L. Kolb

3. Design control is not fully and effectively defined and implemented.
   Action: Revise OP-04 to define an effective and manageable design control process. Target completion June 1996.
   Responsible: D. Jouffroy, S. Specht

4. Corrective action systems for PSD and SNI, internal complaints and customer complaints remain not fully effective ("cumbersome and complicated").
   Action: Contract with K. Hollingsworth to recommend new methods, gain agreement, new systems, and assist implementation for corrective action systems.
   Responsible: Didier Jouffroy

   Software has been written to manage customer complaints. Need training, restoration of network.
   Responsible: S. Bartlett and MIS

   Focus PSD and SNI complaints and corrective actions to Bob Kenny for coordination and resolution.
   Responsible: All

II. Customer Complaints

Strengths

1. CPD system working well.

2. New software development for on-line networking of management.

Weaknesses

1. Further development, refinement and focus required in PSD and SNI.
   Assign to Bob Kenny (PSD, SNI) for focus to track and assure response.
   Responsible: B. Kenny

2. Now need to implement new software system, including status report.
   Responsible: S. Bartlett

III. Customer Returns

Strengths

1. CPD management, tracking, trend charting, and coordination with customers.

2. Database and recording systems in engineering laboratory including failure analysis and reporting has been implemented.

Weaknesses

1. Trend reporting not yet fully implemented for PSD and SNI. Initiate regular report frequency.
   Responsible: J. Lofstrom

2. Untimely analysis, reporting, and closure for portable product returns. Determine analysis and reporting needs and methods.
   Responsible: S. Bartlett and D. Munro

IV. Nonconformances and Corrective Action

Strengths

1. Number of NMRs continues to decrease at incoming inspection.

2. Number of certified parts now exceeds 300.

3. NMR/SCAR data system implemented.

4. CPD continues regular meetings, status review, follow-up, and closeout of issues.

5. Continue to work with target supplier JSA for improvement and Oneida is being replaced. Also replace Triangle Fabricators.

6. Professional purchasing manager hired.

Weaknesses

1. Systems described as cumbersome by K. Hollingsworth have not been changed (see internal audit, item I-1).

2. See item II customer complaints, weaknesses.

3. Quality assurance organization has weakened with departure of key people.
   Action: Reorganize and restaff.
   Responsible: D. Jouffroy and B. Redd

V. Preventive Action

Strengths

1. CPD weekly PA meeting, reporting status, and closure of items continues.

2. Watch program in PSD and SNI with operator control.

Weaknesses

1. PA systems noted as cumbersome by K. Holingsworth and systems not fully effective in PSD and SNI.
   Action: Utilize K. Hollingsworth to recommend, gain acceptance, document, and assist implementation of revised system.
   Responsible: D. Jouffroy

2. Reinforce watch program throughout organization.
   Responsible: J. Kelly, D. Jouffroy

VI. Progress Toward Each Quality Objective

Design and Produce Superior Products

CPD—Reduction of standard deviation of strip thickness.
Action: New band-to-band ductor blade under evaluation. Thickness monitor under evaluation. See attached progress data.

Responsible: P. Bourg

CPD—Reduction of standard deviation of strip hydroxide. Improving trend. See attached.
Action: Continue development.
Responsible: P. Bourg

PSD—Reduction of dielectric leakers on ultrasonic cells.
6/94: 2-40%   10/94: 1-5%   4/95: <1%   10/95: <1%
Goal: 2%
Action: Delete this metric and monitor leakage to maintain control.

Measurably Improve Customer Satisfaction

CPD—Reduction of customer scrap rate.
10/94: 0.75%   3/95: 0.4%   YTD 10/95: 0.91%   Goal: 1.33%

CPD—Improve cycle test yield in Tijuana.
1993: 80%   9/94: 96%   3/95: 98%   9/95: 100%
Goal: 93%

PSD—Reduce late shipments (%). Month end late order analysis.
6/94: 30   9/94: 21   1995 YTD: 24%   10/95: 39%
Goal: 10
Action: Daily status meeting through year end to improve (now in place).

SNI—Reduce late shipments.
Month-end late order analysis.
6/94: 30   9/94: 27   1995 YTD: 11%   10/95: 13%
Goal: 10

Maintain Staff for Low Overhead Organization

CPD—M/wk/employee
366 M/wk/employee total   840 M/wk/direct   Goal: TBD

PSD—Sales per employee (K$)
6/94: 188   10/94: 199   3/95: 215   10/95: See attached.
Goal: +8%/yr.

SNI—Sales per employee
6/94: 356   10/94: 380   3/95: 300   10/95: See attached.
Goal: +8%/yr.

Pursue Continuous Learning and Improvement

CPD—Number of kaizens
6/94: 2   10/94: 2   4/95: 4   10/95: 9   Goal: 10

CPD—Average hours training per employee
4/95: 4.5 Determined status   10/95: 12   Goal: 10
Action: J. Deschenes

CPD—% Personnel passing TAP
In progress

PSD/SNI
Implement team concept.
14 Teams now operating in PSD and SNI   Goal: 14 teams

PSD/SNI—Average hours training per employee
10/95: TBD   Goal: 40
Action: Calculate current status.
Responsible: J. Deschenes

PSD/SNI—% Personnel passing TAP
4/95: 83%   10/95: >83%   Goal: 70%
Action: Determine status.
Responsible: J. Deschenes

On-Time Performance

10/94: 31%   11/94: 42%   12/94: 49%   1/95: 59%
2/95: 57%   3/95: 57%   4/95: 68%   10/95: TBD
Goal: 90%
Action: Further develop system to track customer returns.
Responsible: S. Bartlett

SNI—Improve lead time.

Std. lead time   9/94: 10   4/95: 5   10/95: 5   Goal: 5 days
Exp. lead time   9/94: 5   4/95: 2   10/95: 2   Goal: 2 days

IBD—Customer Survey

PSD/SNI Sales and marketing considering survey system
within customer satisfaction (one of two strategic initiatives
identified at management team meeting, April 28 and 29).
Topic being reviewed for 1996 implementation.
Responsible: S. Bartlett, S. Reames

Maintain Quality Driven Relationships with Customers and Suppliers

PSD / SNI—Charge back suppliers for poor quality. System implemented, but not fully utilized.

CPD—Yearly meeting with six major suppliers.
6/94: 4/6   10/94: 4/6   10/95: 7/6   Goal: 6/6

CPD—Meeting with customers (Nersac and Tijuana).

	10/94	9/95	10/95	Goal
Tijuana	4	2	4/4	4/year
Nersac	6	2	4/4	4/year

VII. Progress Toward Quality Policy

Saft is committed to satisfying our customers with high-value, error-free products and services . . . on time, every time.
Status: See goals in item VI above.

VIII. Organizational Structure, Adequacy of Staffing and Resources

Concern:

1. Competition for resources with multiple high-priority activities and programs continues.

2. Departure of key individuals exacerbates competition for resources noted above.

3. Change in director of quality and management representative is planned for late November.
   Action: New additions to staff: Human resources manager, purchasing manager, director of quality, and offer pending to quality engineer.
   Action: Review organization needs and take corrective action.
   Responsible: B. Redd, J. Deschenes, and staff

IX. Performance, Suitability, and Effectiveness of the Quality System

- ISO 9001 is suitable for our business needs as a foundation quality system.

- Implementation of teams has strengthened our system.

- Watch program fits well within system and strengthens our management system.

- Pursuit of QS-9000 in 1996 is desired, has been committed to customers, but is not underway. Decision in January 1996.

1. Corrective and preventive action system for PSD and SNI require revision to maintain and assure full effectiveness. Plan developed to correct immediate deficiencies and utilize K. Hollingsworth.

2. Design control system revision remains open.

3. More intense management and reporting of the audit system is required.

4. Customer complaints continue to require more focus and management to gain full benefit and effectiveness. Data systems developed and are being debugged for implementation.

5. Trend reporting for customer returns is needed.

6. Quality system is being linked to MFG/PRO for improved total system management.

X. Management Review Summary

1. Focus and intensity is not as high as during precertification (1993–1994).

2. Inadequacies are identified. Some areas have improved. Several items remain open since April management review. See item IX.

3. Competition for resources and departure of key individuals.

4. Quality system is being tied into MFG/PRO systems of key individuals.

5. Introduction of watch program improves operator control. We need to improve this system to maximize this opportunity.

6. Need to start all-employee meetings for improved communication.

CHAPTER

19 Conclusion

ISO 9000 and QS-9000 quality system registration and certification means months of planning, gobs of procedures and work instructions, hundreds of hours of training, lots and lots of hard work and long days, audits and reviews, and let me not forget to mention the money. Lions and tigers and bears, oh no! One thing is for sure: You're not in Kansas anymore.

Here are some things to keep in mind about quality system implementation and third-party registration.

- Don't go for it just to get the framed award.

- Don't do it just for compliance's sake.

- Don't take the plunge because the quality guy wants a new trophy.

- Don't do it because your marketer thinks it would be cool.

- Don't do it in fearful reaction to competition.

- Don't make the commitment just because you want to please your customers (I can hardly believe my own ears).

While all of these reasons have some merit, each pales in comparison to the time, cost, and sacrifice the quality system will take. So why bother? Have you ever heard that quality isn't free, but it pays? For a moment, reflect on what you've read in previous chapters. The task (an understatement) of developing, implementing, registering, and maintaining your quality system, if done very well, will teach you and your employees more about the guts of the business than perhaps any other approach. Jobs will be better and more thoroughly understood. Processes (administrative *and* operational) will be examined and refined as never before. Exercising the system over time will promote consistency and will drive improvement in everything from entering an order to shipping the widget out the door. Any activity (service, administration, manufacturing, and so forth) must be brought into a steady state and be under control before it can be improved. Your quality system will help reduce variation and minimize anomalies.

The system will lead you away from fire-fighting and out of the mire of never-ending fix-it loops. It can steer you toward truly proactive, preventive stuff (even if some folks come kicking and screaming at first). Isn't it interesting how certain people can function only in the crisis management mode? The system can expose redundant, wasteful, ineffective, and inefficient expenditures of valuable resources. It can condition and prepare the company to not only accommodate change, but to manage change. Its quality policy will help to maintain constancy of purpose and motivation. The measurable quality objectives will provide genuine performance indicators without the smoke and mirrors.

The ISO 9000 series standards are here to stay. They have been adopted by more than 100 countries. More than 100,000 companies have been registered worldwide, of which my company has registered more than 10,000. The Big Three automakers (and major truck manufacturers as well) have adopted verbatim the ISO 9001 standard into their own requirements, QS-9000. As the U.S. military continues to phase out the MIL documents, ISO standards are the preferred choice to fill the vacuum. They are usually the model of quality systems for companies seeking CE Marking (for product export to Europe) and for ISO 14000 (environmental management system registration).

Select your registrar early. Select one that is internationally known, globally located, well accredited, mature, and experienced with your type of business. Select one that subscribes to a partnering, value-adding relationship with its customers. Select one that almost exclusively uses its own trained and qualified employees, rather than relying heavily on subcontracted assessors. Select a registrar that represents the greatest value—one that is not necessarily the least expensive. While consultants, educators, and technical user groups have a place, be sure to understand your own registrar's expectations and interpretations. The registrar will make the final evaluation and eventual registration of your system.

Is this a panacea, a miracle, a cure-all? No! But it's one of the very best tools for business management and improvement that I've been involved with during the past 15 years. Remember quality circles? They were good, but not great. Still scratching your head over TQM? It could have been great if ISO 9000 or QS-9000 implementation had been a prerequisite.

Here are some reasons for quality system implementation and third-party registration.

- Do it to increase productivity.

- Do it to develop your employees.

- Do it to control all your processes.

- Do it to manage inevitable change.

- Do it to improve product/service quality.

- Do it to improve performance results.

- Do it to get rid of the smoke and mirrors.

- Do it to increase effectiveness and efficiency.

- Do it because it makes sound business sense.

- Do it because you know in your gut it's the right thing.

- Do it because Simon says!

Index

Internal quality audits
    assessment readiness and, 20, 23, 28
    management review of, 163–64
    as processes, 132
    records of, 28
Internal relations, agreements on, 139–44
International Organization for Standard-
    ization (ISO), ix, xiv. *See also specific*
    ISO standards
Interpretations, by registrars, 27, 101
Interrelations
    change management and, 52
    interface agreements and, 139–44
ISO 9000 series requirements standards, xiv
    paragraph 1., 26–27
    paragraph 4.1.1., 58
    paragraph 4.1.2.3., 1
    paragraph 4.1.3., 64
    paragraph 4.1.4., 64
    paragraph 4.2.1., 117
    paragraph 4.3., 77, 87–88
    paragraph 4.3.2., 79–80
    paragraph 4.3.3., 81
    paragraph 4.3.4., 82
    paragraph 4.6., 88
    paragraph 4.9., 88
    paragraph 4.11., 89
    paragraph 4.12., 25
    paragraph 4.14.2., 23
    paragraph 4.14.3., 23
    paragraph 4.15., 89
    paragraph 4.16., 27, 64, 82, 89–90
    paragraph 4.17., 20, 23
    paragraph 4.18., 39, 90
    paragraph 4.20., 90–91
    paragraph 4.20.1., 146
    paragraph 4.20.2., 146
    plant services and, 86–91
    as platform for change, 49–56
    spirit of, 24, 26–27
    whole organization and, 85
    worldwide acceptance of, 174
ISO 9001, xiv, 1, 77, 145
ISO 9002, 77, 145
ISO 10013, xiv, 11, 117
ISO 14000 series, 97, 174

M

Management representative
    authority and responsibility of, 1–7
    corrective action and, 126
    management review and, 72

    in multisite environments, 141
    quality manual and, 118
Management reviews, 63–75
    attendance roster for, 70–71
    attributes of, 33
    change management and, 51–52
    effectiveness and, 66–67, 68–69, 170
    frequency of, 71–72, 161
    importance of, 63–64
    individuality of, 69–70, 162
    inputs to, 72–74
    outputs from, 74–75
    process of, 64, 69–75, 132
    purpose of, 64–65
    quality objectives and, 58
    questions for, 64, 67–68
    at Saft America, 161–71
    suggestions for, 162
    suitability and, 65–66
    summary example for, 163–71
Master document/data lists, 108
Matrix, of skills, 46
Maturity
    of registrars, 95–96
    of systems, 31–35
Measurements, of effectiveness, 68–69. *See
    also* Indicators
Memos of understanding (MOUs), 139.
    See also Interface agreements
Multisite environments, 140–41

N

Naming schemes, for documents and data,
    107
Needs identification
    for statistical techniques, 145–52
    for training, 41, 44–45
Nonconformances
    clearing of, 102, 123–29
    exceptional, 128–29
    management review of, 165–66
    quantity of, 36
Number schemes, for documents and data,
    107

O

Objective evidence, 26–28, 35–37
    of corrective action, 126–27
    of training, 41, 47
    types of, 36–37

On-the-job training, 46
On-time performance, 168
Organizational goals, 60
Organizations
    interface within, 139–44, 156
    multisite, 140–41
    peripheral units of, 85–86
    quality throughout, 40, 85–86
    structure of, 169

P

Packaging, in plant services, 89
Performance, of quality systems, 170. *See
    also* Effectiveness
Peripheral units, 85–86
Personalization. *See* Customization
Personnel. *See also* Training
    for administrative processes, 133–34
    for contract reviews, 78–79, 81–82
    for design control, 156, 157
    for document/data review, 108
    efficiency and, 30
    for management reviews, 70–71,
        163–69
    positions listed for, 40–41, 42, 43
    skills matrix for, 46
    for statistical techniques, 150
Planned arrangements, 20–21
Plant services, 85–91
Pre-assessments, 102, 153–54
Predictability, and statistical techniques,
    146
Predictable outputs, 24
Preservation, in plant services, 89
Preventive action, 23, 54, 166
Price, and the registrar
    by components, 113
    by quality factors, 114, 115
    questions and answers on, 99–101
    value versus, 111–15
Prime intents, 27
Proactive opportunities, 54
Process, definition of, 69
Processes
    administrative, 35, 131–37
    capability of, 55, 149
    control of, 88, 149
    definition of, 132
    of management review, 64, 69–75
    special, 41
Procurement, and change management, 55
Production part approval, 159

Product quality, and change management,
    55
Project plans, 156, 157, 158
Protection, of documents and data,
    108–09
Prototypes, 158
Purchase orders, 80–81
Purchasing
    change management and, 55
    in multisite environments, 140–41
    in plant services, 88
    as a process, 132

Q

QS-9000, xiv
    acceptance of, 174
    contract reviews and, 77
    improvement and, 32
    plant services and, 86–91
    as platform for change, 49–56
    registrar quality system and, 97
    statistical techniques and, 145
    training and, 42
    whole organization and, 85
QSN (quality system nonconformity)
    forms, 125–29. *See also* Corrective
    action
Qualifications, and statistical techniques,
    146
Quality
    throughout the organization, 40,
        85–86
    value, price, and, 111–15
Quality assurance manual. *See* Quality
    manual
Quality commitment statement, 51
Quality manual, 11–13
    audit of, 11, 13, 36, 117–21
    changes in, 12, 119
    customization of, 119
    design control and, 155
    minimum requirements for, 117–19
    purpose of, 11
Quality objectives
    attributes of, 58–59, 60–62
    change management and, 51
    examples of, 58, 59–60
    management review of, 166–69
    quantity of, 60
Quality planning
    change management and, 52–53
    as a process, 132